SpringerBriefs in Environmental Science

SpringerBriefs in Environmental Science present concise summaries of cutting-edge research and practical applications across a wide spectrum of environmental fields, with fast turnaround time to publication. Featuring compact volumes of 50 to 125 pages, the series covers a range of content from professional to academic. Monographs of new material are considered for the SpringerBriefs in Environmental Science series.

Typical topics might include: a timely report of state-of-the-art analytical techniques, a bridge between new research results, as published in journal articles and a contextual literature review, a snapshot of a hot or emerging topic, an in-depth case study or technical example, a presentation of core concepts that students must understand in order to make independent contributions, best practices or protocols to be followed, a series of short case studies/debates highlighting a specific angle.

SpringerBriefs in Environmental Science allow authors to present their ideas and readers to absorb them with minimal time investment. Both solicited and unsolicited manuscripts are considered for publication.

Moonisa Aslam Dervash • Abrar Yousuf
Munir Ozturk • Rouf Ahmad Bhat

Phytosequestration

Strategies for Mitigation of Aerial Carbon
Dioxide and Aquatic Nutrient Pollution

Moonisa Aslam Dervash
Division of Environmental Sciences
Shere Kashmir University of Agricultural
Sciences and Technology of Kashmir
Srinagar, India

Munir Ozturk
Faculty of Science
Ege University
Izmir, Turkiye

Abrar Yousuf
Regional Research Station
Punjab Agricultural University
Ludhiana, Punjab, India

Rouf Ahmad Bhat
Division of Environmental Sciences
Sher-e-Kashmir University of Agricultural
Sciences and Technology, Kashmir
Srinagar, Jammu and Kashmir, India

ISSN 2191-5547   ISSN 2191-5555   (electronic)
SpringerBriefs in Environmental Science
ISBN 978-3-031-26920-2   ISBN 978-3-031-26921-9   (eBook)
https://doi.org/10.1007/978-3-031-26921-9

This Springer imprint is published by the registered company Springer Nature Switzerland AG
The registered company address is: Gewerbestrasse 11, 6330 Cham, Switzerland

*Dedicated to "THE SAFFRON TRUST"
(London, United Kingdom) for their
services in terms of "relief of financial
hardships, alleviation of health problems and
advancement of education to the people of
Jammu and Kashmir (India)." A great
appreciation for their selfless services
towards the betterment of humanity.*

Preface

Climate change and cultural eutrophication are among the most imperative environmental concerns across the world. The temperature escalation has now been explicitly verified and continues with an unprecedented pace. The apprehensions of global warming include retreat of glacial resources, submergence of low-lying areas, and outbreak of various diseases, thus impacting the existence of mankind coupled with extermination of flora and fauna which can lead to ecological imbalances. On the other hand, cultural eutrophication due to the varied array of connotations pertaining to industrialization and urbanization has led to proliferation of nuisance weeds in aquatic environs which disturb the ecological status of such ecosystems.

Greenhouse effect is the natural phenomena prevailing on Earth which aids in optimizing its average temperature, but an alarming increase in the concentration of such gases (CO_2, CH_4, CFCs, N_2O, and water vapor) into the atmosphere causes enhanced greenhouse effect which is responsible for global warming. Among all the greenhouse gases, CO_2 is accountable for causing 55–60 percent of the total enhanced greenhouse effect. In order to curb the catastrophic consequences of global warming, carbon sequestration by plants is a viable cost-effective tool to mitigate ever-growing concentration of CO_2 from the air. Such a sequestration through plants is the absorption and long-term storage of atmospheric CO_2 by the plant cover, besides uptake. Plants absorb CO_2 from the atmosphere to carry out food manufacturing process (photosynthesis) and finally incorporate it into biomass.

On the other hand, aquatic contamination due to nutrient enrichment is a global concern which is the aftermath of domestic and agricultural runoff. In the present era of industrialization and urbanization, terrestrial as well as aquatic environs are heavily loaded with contaminants like CO_2, nitrogen (N), and phosphorous (P), which lead to proliferation of noxious macrophytes. There are chemical, physical, and biological techniques available for remediation of nutrients, but in addition to carbon sequestration by plants, nutrient sequestration via plants is an ecofriendly approach which can be applied to get rid of such contaminants. Therefore, in order to sequester excess CO_2 from the air and excess nutrients from the contaminated environs, carbon-nutrient sequestration can be braided together as phytosequestration to achieve goals of sustainable development.

The present book proposal is a unique assortment of the two major environmental concerns together in a single volume and various chapters are meant to bring forth the historical perspective of global warming, sources and monitoring tools for greenhouse gases, impacts of temperature escalation, carbon capture and storage, and climate models, together with the societal responses to climate change. As a follow-up, the succeeding chapters enlighten the monitoring of nutrient pollution, ecological impacts of nutrient contamination vis-a-vis phytomediation of nuisance pollution.

The present book will serve as a brief and brisk reference for academicians, researchers, and students regarding the climate change and nutrient pollution with diverse concepts focussing monitoring strategies, consequences of pollution escalation and mitigative measures. Therefore, it will be an adequate and coherent bind for meeting the diverse requirements of teaching and research.

Srinagar, India   Moonisa Aslam Dervash
Ludhiana, Punjab, India   Abrar Yousuf
Izmir, Turkiye   Munir Ozturk
Srinagar, Jammu and Kashmir, India   Rouf Ahmad Bhat

Acknowledgment

This manuscript has been written as a collective effort by the authors mentioned in this book. The authors are thankful to and would like to convey their gratitude to the "SPRINGER TEAM" for their continuous support, patience, and friendly collaboration.

Contents

About the Authors

Moonisa Aslam Dervash, Ph.D., has obtained her PhD from Division of Environmental Sciences, SKUAST-K, India. Her specialization is in environmental monitoring, environmental awareness, ecology, soil biology, wetland restoration, and carbon sequestration. She has authored more than 6 books with International publishers. Moreover, she has published more than 35 scientific research articles and book chapters in the journals of national & international repute. She has also participated and presented her papers in national and international scientific conferences. She is recipient of many prestigious awards and felicitations for her dedicated accomplishments. Her focus has remained on many facets of society especially on environmental conservation and women empowerment. She has been felicitated by state government (J & K Department of Ecology, Environment and Remote Sensing) for her outstanding contribution in mass environmental awareness and conservation through electronic media (All India Radio, Srinagar). Recently, she has been awarded Postdoctoral Fellowship by Ministry of Education (ICSSR), Government of India.

Abrar Yousuf, Ph.D., is working as a Scientist (Soil and Water Engineering) at RRS Ballowal Saunkhri, Punjab Agricultural University, Punjab. His field of specialization is watershed hydrology, soil erosion modeling, watershed management, remote sensing and GIS, rainwater management, and its judicious use. He is involved in continuous monitoring of runoff and sediment yield from the various watersheds located in Kandi region of Punjab. He has been involved in four research projects funded by various funding institutes including ICAR New Delhi, DST New Delhi, GIZ New Delhi, and IPROconsult Germany. Currently, he is handling two research projects as principal investigator which are being funded by the Department of Soil and Water Conservation, Government of Punjab. He is working on ex-situ management of rainwater in farm ponds and its judicious use through micro-irrigation systems. He has constructed farm ponds in various adopted villages including Behdarya (ORP village), Kambala (RIFS village), and Kangar (SCSP village). He has authored a number of technical bulletins, a book, and many research articles in journals of national and international repute. He is a member of many scientific associations and actively participates in scientific conferences, workshops, summer and winter schools, training programs for building up high scientific temper and technical skills. He has many reputed publications including research articles, book, book chapters, and technical articles to his credit. Recently, he has been felicitated with "Best Thesis Award" for his doctoral research work by Soil Conservation Society of India, New Delhi.

Munir Ozturk, Ph.D., D.Sc., has served at the Ege University-Turkiye for more than 50 years in different positions and has been Founder Director of the Centre for Environmental Studies, Ege University, and Chairman of the Botany Department as well as Director of the Botanical Garden. *Sideritiso zturkii* and *Verbascum ozturkii* are 2 newly recorded endemic plant species from Turkiye in the name of Munir Ozturk. His fields of scientific interest are pollution and biomonitoring, plant eco-physiology, biosaline agriculture, and medicinal-aromatic plants conservation. Dr. Ozturk has published almost 60 books with internationally known publishers including Springer, Elsevier, Taylor & Francis, Wiley, and few others. He has more than 80 book chapters and 200 papers in international journals; 120 with impact factor. Dr. Ozturk has also presented 125 papers at the international meetings and 85 at the national meetings and has served as a guest editor for more than 13 international journals. He holds many Memberships of "Institutions and Professional Bodies" and has received fellowships from the globally recognized Alexander von Humboldt Foundation, Japanese Society for Promotion of Science, and the National Science Foundation of the USA. He has also worked as consultant fellow at the Faculty of Forestry, Universiti Putra Malaysia, Malaysia; as Distinguished Visiting Scientist at International Centre for Chemical and Biological Sciences, ICCBS-TWAS, Karachi University, Pakistan; as "Vice President of the Islamic World Academy of Sciences" from 2017 to 2022; and is a "Fellow of the Islamic World Academy of Science" as well as "Foreign Fellow Pakistan Academy of Science."

Rouf Ahmad Bhat, Ph.D., has pursued his doctorate at Sher-e-Kashmir University of Agricultural Sciences and Technology Kashmir (Division of Environmental Science) and presently working in the Department of School Education, Government of Jammu and Kashmir. Dr. Bhat has been teaching graduate and postgraduate students of Environmental Sciences for the past three years. He is an author of more than 50 research articles (h-index 25; i-index 35; total citation >2000) and 40 book chapters and has published more than 35 books with international publishers (Springer, Elsevier, CRC Press Taylor and Francis, Apple Academic Press, John Wiley, and IGI Global). He has specialization in Limnology, Toxicology, Phytochemistry, and Phytoremediation. Dr. Bhat has presented and participated in numerous State, National, and International conferences, seminars, workshops, and symposiums. Besides, he has worked as an associate Environmental Expert in

World Bank-funded Flood Recovery Project and also as the Environmental Support Staff in the Asian Development Bank (ADB) funded development projects. He has received many awards, appreciation, and recognition for his services to the science of water testing, and air and noise analysis. He has served as an editorial board member and reviewer of reputed international journals. Dr. Bhat is still writing and experimenting with diverse capacities of plants for use in aquatic pollution remediation.

Chapter 1
Introduction to Phytosequestration: Strategies for Mitigation of Aerial Carbon Dioxide and Aquatic Nutrient Pollution

Carbon is one of the most essential components of biological world and the associated environment (physical attributes). From the time immemorial, carbon has occupied the forefront position in operating a varied array of functions ranging from "cellular level (e.g., precursor in metabolic processes) to ecosystem functioning (e.g., temperature regulation via greenhouse effect)." Greenhouse gases act as a circuit house to regulate the temperature of Earth at an optimum of 15 °C by trapping the infrared wavelengths. If greenhouse gases were not present in the atmosphere, the temperature of Earth would have been ranging in sub-zeros (−18 °C to −20 °C).

As per data retrieved from National Oceanic and Atmospheric Administration (NOAA 2019), the present level of CO_2 in the atmosphere has crossed 403.64 ppm compared to the concentration at the onset of the Industrial Revolution. The main proportion of CO_2 (i.e., 75%) stems out from fossil fuel burning, whereas 25% is due to obliteration of tropical rain forests (Ozturk et al. 2015a). The intensification of CO_2 is envisaged to cause significant alteration in climatic conditions which eventually leads to climate change. Intergovernmental Panel on Climate Change in 2018 revealed that anthropogenic activities are estimated to have caused approximately 1.0 °C of global warming above pre-industrial levels, with a likely range of 0.8–1.2 °C and is likely to reach 1.5 °C during 2030 to 2052 (IPCC 2018). If escalation of GHGs in the atmosphere pacifies at the existing rate, it will cause rise of the global sea level by 0.26–0.77 m at the end of the twenty-first century due to the alleviation of polar ice, which would badly impact low-lying coastal areas, existence of various plants and animals, agriculture industry, and livelihood patterns, and eventually there would be an outbreak of various diseases like dengue, malaria, etc. Consequently, the accumulation of greenhouse gases in the atmosphere, particularly CO_2, has amplified; therefore the multiple mitigation measures to stabilize or reduce its concentrations need to be addressed through a hierarchy of actions with most important being bio-fixation of carbon as a part of the comprehensive strategy based on the principle of removing much of CO_2 from the atmosphere.

© The Author(s), under exclusive license to Springer Nature Switzerland AG 2023
M. A. Dervash et al., *Phytosequestration*, SpringerBriefs in Environmental
Science, https://doi.org/10.1007/978-3-031-26921-9_1

Climate system makes our planet Earth a habitable entity in the Universe. The climate system changes over time as a result of both internal dynamics like volcanic eruptions and changes in external factors like solar radiation and air composition. It is one of the oldest systems of the Earth that has significantly contributed to the origin and evolution of life through ages (Ozturk et al. 2015a). The present climate can be deciphered by studying the components of climate system. Deciphering the past and present climate requires the use of instrumental records and environmental indicators. The human population is one of the main drivers of ecological footprints in the current climate. The disproportionate population explosion made it much harder to control global warming and resource depletion. The industrialization, deforestation, urbanization, desertification, and stratospheric ozone depletion, which are the major issues faced globally in this twenty-first century, are due to the ever-growing human population. Scientists all over the world are thinking of various strategies to understand the phenomenon of climate change, to gain an insight about its implications, and also to quantify the changing climate by means of models. These models attempt to quantify the changing climate by using physical laws of radiation and energy and studying the radiation behavior and flux at the surface of the Earth. So, there may be regional or global climate models to quantify the extent of change that the Earth is undergoing in terms of climate and also include clouds and aerosols in more complex models for understanding their dual role in radiative forcing of the Earth that eventually leads to climate change. Various emission scenarios of greenhouse gases are also discussed in this book. A comprehensive account of past climate which deals with the climate of the past, sources of past climatic conditions, and changes in climate during the Quaternary Period, when humans appeared as a dominant biotic element on the Earth, is also discussed in detail.

On the other hand, water that has been contaminated by an excessive amount of nutrients is known as nutrient pollution. It is a major factor in the eutrophication of surface waterways, such as lakes, rivers, and coastal waters, where an overabundance of nutrients, typically nitrogen or phosphorus, encourages the growth of algae. Surface runoffs from agricultural fields and pastures, septic tank and feedlot discharges, and combustion emissions are all sources of nutrient pollution. Due of its high nutrient content, untreated sewage plays a significant role in cultural eutrophication. Numerous serious environmental issues are connected to an overabundance of reactive nitrogen molecules in the environment. These include accelerated eutrophication, proliferation of algal blooms, hypoxia, acid rain, nitrogen saturation in forests, and climate change.

This book is a unique assortment of carbon dioxide, nitrogen, and phosphorous which are limiting factors for the survival of life on Earth up to an optimum threshold. Once the benchmark of threshold limit exceeds, these limiting factors transform into noxious pollutants. Therefore, it is essential to highlight the sources, monitoring, and consequences. Finally, this Springer brief also focuses on mitigation strategies of these emerging pollutants through carbon sequestration and phytoremediation.

Fig. 1.1 Sequestration of atmospheric CO_2 and aquatic nutrient pollution

Carbon sequestration through vegetation and soil is the process of capture and long-term storage of atmospheric CO_2 which seems to be an economical and feasible option to mitigate the increasing concentrations of CO_2 (Fig. 1.1), whereas a collection of procedures known as phytoremediation employ plants to reduce, eliminate, degrade, or immobilize environmental toxins, primarily those with anthropogenic origins, with the goal of rehabilitating nearby places so they can be used for either private or public purposes (Hakeem et al. 2015; Ozturk et al. 2015b, c). The main goals of phytoremediation have been to remove dangerous heavy metals from soils or water or to employ plants to speed up the decomposition of organic pollutants, usually in collaboration with rhizosphere bacteria. Besides, macrophytes have been extensively utilized for sequestration of nutrients from contaminated aquatic environs. In this bind, seven features of phytoremediation are briefly discussed in the last chapter: rhizosphere degradation, rhizofiltration, phytoextraction, phytodegradation, phytostabilization, phytovolatilization, and phytorestoration (Hakeem et al. 2015; Ozturk et al. 2015b, c).

This book is a thoughtful effort in bringing forth carbon sequestration and phytoremediation (Hakeem et al. 2015; Ozturk et al. 2015b, c). The opening chapters of the manuscript deal with comprehending the historical outlook of global warming highlighting the climate of the past and present, sources and monitoring tools for greenhouse gases in the atmosphere, consequences of global warming, carbon capture and storage techniques, climate models for prediction of future climate, and societal responses to climate change. The book's last chapters have been devoted to the monitoring of nutrient pollution in water, ecological impacts of nutrient contamination, and phytoremediation of nutrient pollution.

References

Hakeem KR, Sabir M, Ozturk M, Mermut A (2015) Soil remediation and plants: prospects and challenges. Academic Press, Elsevier, London, 724 pp

IPCC (2018) Framing and context. Allen MR, Dube OP, Solecki W, Aragon-Durand F, Cramer W, Humphreys S, Kainuma M, Kala J, Mahowald N, Mulugetta Y, Perez R, Wairiu M, Zickfeld K. In: Global warming of 1. 5°C. An IPCC Special Report on the impacts of global warming of 1.5°C above pre-industrial levels and related global greenhouse gas emission pathways, in the context of strengthening the global response to the threat of climate change, sustainable development and efforts to eradicate poverty (Eds. Delmotte VM, Zhai P, Pörtner HO, Roberts D, Skea J, Shukla PR, Pirani A, Okia WM, Péan C, Pidcock R, Connors S, Matthews JBR, Chen Y, Zhou X, Gomis MI, Lonnoy E, Maycock T, Tignor M, Waterfield T)

NOAA (National Oceanic and Atmospheric Administration) (2019) Atmospheric CO_2 Mauna Loa Observatory (Scripps / NOAA / ESRL). Monthly & annual mean CO_2 concentrations (ppm). Washington, DC

Ozturk M, Hakeem KR, Faridah-Hanum I, Efe R (2015a) Climate change impacts on high-altitude ecosystems. Springer Science+Business Media, XVII, New York, 695 pp

Ozturk M, Ashraf M, Aksoy A, Ahmad MSA (2015b) Phytoremediation for green energy. Springer Science+Business Media, New York, 191 pp

Ozturk M, Ashraf M, Aksoy A, Ahmad MSA (2015c) Plants, pollutants & remediation. Springer Science+Business Media, New York, 407 pp

Chapter 2
Global Climate: Chronological Perspective

2.1 Introduction

The climate is the average weather over a long period of time that affects a vast area, possibly the entire globe. The state of the atmosphere over a specific area is represented by the weather on an hourly or daily basis. The major elements of the climate include the atmosphere, ocean, snow and ice cover, land surface, rotation and revolution of the globe, incoming solar radiation, and biota. Their interaction creates a system known as the climate system. In fact, it is the climate system, which makes our planet Earth a habitable entity in the Universe (Ozturk et al. 2015). The climate system varies over time as a result of changes in external factors like solar radiation and air composition as well as its own internal dynamics like volcanic eruptions. It is one of the planet's oldest systems and has made a substantial contribution to the beginning and progression of life across time. Since the Earth's formation approximately 4.6 billion years ago, it has undergone continuous change, and as a result, the planet's climate has also undergone periodic alteration. By examining the elements that make up the climate system, we may learn about the current climate. However, knowledge of both the present and the past climates is required to comprehend the current dynamics of the climate and anticipate its future condition. The past climate, the factors that influenced historical climatic conditions and climate changes during the Quaternary Period, when humans emerged as the planet's main biotic element, is all depicted in detail in this chapter.

© The Author(s), under exclusive license to Springer Nature Switzerland AG 2023
M. A. Dervash et al., *Phytosequestration*, SpringerBriefs in Environmental
Science, https://doi.org/10.1007/978-3-031-26921-9_2

2.2   Paleoclimate

The past climate is referred to as paleoclimate, whereas the science of examining the contemporary climate is known as climatology. Thus, paleoclimatology is the field of study that deals with the analysis of past climatic conditions. The word paleoclimatology is a combination of the Greek words *Palaios* (ancient) + *clima* (climate) + *ology* (branch of learning), and therefore it refers to the study of the past climate. Paleoclimatologists are the specialists who research past climates. Before there were available records of instrumental climate data, paleoclimatologists used natural environmental evidence or their proxies found on the surface of the Earth, such as sediments, sedimentary layers, fossils (such as the growth rings of coral and trees), ice cores, and radiocarbon, to infer the climate of the past.

It is startling to learn that climatic fluctuations are neither an uncommon nor a novel event. Climate change is a natural phenomenon, and since the planet's creation, it has gone through several cycles. Let's get acquainted with the geologic time scale before talking about the climate of the world in the past. Geological time, which spans the entire Earth, is split into specific geologic time units like decade, era, period, epoch, and age. These geologic time units are analogous to the divisions of our time into years, months, weeks, days, hours, minutes, and seconds (Table 2.1). By dating rocks with radioactive techniques, it is possible to ascertain the length of a specific time unit on the scale. The boundary between two time units is typically marked by abrupt biotic events, such as the origin or extinction of a particular species.

2.3   Glimpse of the Earth's Climate Through Ages

Numerous evidences about former climates may be found in the rock record of the Earth, demonstrating that the climate of the living planet Earth has not been constant from its beginnings almost 4.6 billion years ago to the present. The main sources of proxy data for paleoclimate are fossils (tree rings, plant leaves, pollen, and coral skeletons), ice cores, sedimentary strata, and sediment. The analysis of these proxy data also shows that solar activity, volcanic eruptions, lithospheric plate movements, weathering reactions, as well as variations in greenhouse gases and temperatures frequently changed the Earth's climate in the past. Additionally, cyclical variations in the Earth's orbit around the Sun and changes in oceanic circulation patterns owed to the Earth's climate. The Earth's ancient climate was also impacted by biological evolution and extraterrestrial (meteorite) effects. It should be emphasized that non-glacial (intra- and inter-glacial) and glacial periods are used to identify the past climate of the planet (Evenick 2021).

Table 2.1 Summary of the geological time scale showing main time units

Eon	Era	Period	Epoch	Time interval in million years (Ma)
Phanerozoic	Cenozoic	Quaternary	Holocene	0.012 to present
			Pleistocene	2.58 to 0.012
		Neogene	Pliocene	5.333 to 2.58
			Miocene	23.03 to 5.333
		Paleogene	Oligocene	33.9 to 23.03
			Eocene	56 to 33.9
			Paleocene	66 to 56
	Mesozoic	Cretaceous	–	145 to 66
		Jurassic	–	201.3 to 145
		Triassic	–	251.9 to 201.3
	Paleozoic	Permian	–	298.9 to 251.9
		Carboniferous	–	358.9 to 298.9
		Devonian	–	419.2 to 358.9
		Silurian	–	445.2 to 419.2
		Ordovician	–	485.4 to 445.2
		Cambrian	–	541 to 485.4
Precambrian	Proterozoic	Neoproterozoic	–	1000 to 541
		Mesoproterozoic	–	1600 to 1000
		Paleoproterozoic	–	2500 to 1600
	Archean	Neoarchean	–	2800 to 2500
		Mesoarchean	–	3200 to 2800
		Paleoarchean	–	3600 to 3200
		Eoarchean	–	4000 to 3600
	Hadean	–	–	4600 to 4000

2.3.1 Climate During Precambrian

Climate on Earth was warm, and concentrations of greenhouse gases including carbon dioxide, methane, and water vapor were very high during the Precambrian period (4.6 billion to 540 million years ago). The amount of methane was above 1000 ppm, and the concentration of carbon dioxide (CO_2) was more than 18 times higher than it is now. In the early atmosphere, oxygen wasn't present. The Earth's temperature dropped after millions of years of planet formation, and rain was produced by the condensation of water vapor in the early atmosphere. As a result, the world was endowed with fundamental necessities including soil, water, and air for the beginning of life. The earliest life forms, including cyanobacteria, initially appeared on the Earth's surface some 3.5 billion years ago. Around 600 million years ago, these bacteria produced their own food by using sunlight as a source of energy and releasing oxygen as a byproduct of photosynthesis. The atmosphere contained enough oxygen for multicellular life to develop. In the Earth's Precambrian history, both the non-glacial and glacial periods have been documented. Evidence from sedimentary rocks left behind by glaciers indicates that Earth was extremely cold, perhaps even close to freezing point, during the Precambrian period (Ruddiman 2001). From the Precambrian period, four ice ages are documented: the first one occurred in the Archean eon about 2500 million years ago, and three others occurred in the Proterozoic era between 900 and 600 million years ago (Barry and Chorley 2010).

2.3.2   Climate During Phanerozoic

The amount of carbon dioxide in the atmosphere changed dramatically during the Phanerozoic eon (540 million years ago to the present), dropping from 6000 ppm to its present levels. The Phanerozoic climate was significantly influenced by the carbon cycle, which led to a wide range of variation in multicellular creatures and land plants (Beerling and Berner 2005). During the Phanerozoic period, it has been observed that the climate frequently fluctuated between icehouse (glacial and non-glacial) and greenhouse conditions, with temperatures being greatly influenced by natural processes such as the breaking and reunification of continental landmasses as well as extraterrestrial impacts. The Phanerozoic history of life contains records of the five main mass extinctions, including the End Ordovician, End Devonian, Permian/Triassic boundary, End Triassic, and Cretaceous/Tertiary boundary. The widespread changes to the past climate are directly responsible for all of these catastrophic extinctions. The rapid and irreversible disappearance of a significant number of species or groups of creatures from the surface of the world is known as mass extinction. The Phanerozoic eon, which began in the Ordovician period and ended in the Late Cenozoic era, is known to have seen three significant ice ages (Barry and Chorley 2010).

2.4   Sources of Paleoclimate Data

It should be mentioned that climatologists, also known as climate scientists, can access data from a century ago in order to understand the contemporary climate. Can you believe that this information, which is only about 4.6 billion years old, is enough to know the climate of the entire Earth? Without a doubt, the answer is no. In order to reconstruct the Earth's past climate, paleoclimatologists used climatic archives or proxies. The climatic archives are made up of the Earth's materials as well as ancient written records (such as historical records) that hold the physical traits of the past environment. The past's climate can be recreated by looking through climate records. Historical data, archeological data, and geological records are the three basic categories of climate archives.

2.4.1 Historical Data

For retracing the historical climate, it serves as the first source of data. Documentary data makes up this information. The sources of historical information include farmer's diaries, traveler's diaries, newspapers, paintings, artistic representations, reports from early weather observers, and other public records. Aside from these, there are other essential sources of information for reconstructing past/ancient climates, including legal documents, written accounts, tax records, economic records, and visual records, that offer information about land uses, landscapes, societal collapse, construction materials, and biodiversity. A record of the date of forests and tree blooming; the incidence of snowfall, rain, drought, famine, and flood; as well as the migration of birds is also considered historical data. Historical records offer information about the past climate that is both qualitative and quantitative. The Mesopotamian culture of the Middle East, for instance, was thought to be one of the first civilizations to record events, and these data give climatic details of the events as documented by the humans.

2.4.2 Archeological Data

Archeology deals with the study of the past human cultures. It focuses on how people in the past moved, worked, traded, and lived. Archeologists are professionals in the field of archeology. To understand how climatic conditions affect prehistoric humans' way of life, researchers study archeological data. It should be mentioned that archeological data is significantly older than historical data since it represents a period of ancient cultures that have been recreated by scientific examination of multiple soil layers that have preserved human artifacts. In other words, unlike historical data, which is based on records made by humans, archeological time is based on the remains of human life (Krebs 2007). An archeological site is a location where remnants of ancient human activity have been preserved. These remnants are a valuable source of knowledge on both cultural and non-cultural aspects of the environment, which together make up the archeological data (Reitz et al. 2008). Environmental archeology is the branch of archeology that focuses on reconstructing the past environment, particularly climate. The following types of information are retrieved from an archeological site that could be helpful for paleoclimatic research: rock layers, minerals, and soil data.

Numerous clues about the historical climate can be found in the chemical, physical, and geological features of soil, rock layers, and mineral samples collected from any location. The investigations of sediment layer grain sizes provide information on the deposition medium (wind, water, floods, or glaciers), which is helpful in reconstructing past climatic conditions. The sequence of sediment layers of nearshore archeological sites gives information of the sea-level changes as they contain distinct layers of sediments deposited under marine to freshwater conditions.

2.4.2.1 Plant and Animal Remains

Any kinds of plant material, including wood, mature seeds, pollen, spores, fruits, flowers, leaves, stems, roots, bark, epidermis, fibers, stomata, starch grains, phytoliths, resins, lignin, and lipids associated with any place, are included in the plant remains. Mammal bones and teeth, fish skeletons, and invertebrate shells make up the animal remnants (mollusks, echinoderms, crustaceans, insects, foraminifers, and protozoans). It is a proven fact that climatic factors heavily influence plant and animal life. As a result, studying their remnants will enable us to identify their food source and recreate historical climatic and environmental circumstances. An understanding of paleodiets (terrestrial/aquatic diets), paleotemperature (i.e., past temperatures), and seasonal patterns is provided by the carbon, nitrogen, and oxygen isotopic studies of bone and shell remains. The existence of mammoth (elephant) remnants made up of bones and skins plainly indicates a cold climate. The presence of floral remains largely consisting of bouquets of wild flowers in the burial sites provides information about climate conditions prevailed at the time the society lived there. A site containing ancient humans and domesticated animals was abruptly naturally buried, which may have pointed to the flooding phenomenon, which is connected to rainfall and provides crucial hints about the previous climate.

2.4.2.2 Artifacts

These are objects created, modified, and used by humans. These are things that people have made, changed, and used. Artifacts include different types of pottery (intact or broken); tools made of stone, wood, bone, and metal (arrowheads, maceheads, and spears); ornamental items (jewelry and figurines or statuettes); and clothes. It is believed that prehistoric individuals produced or created the items using locally accessible materials. So their research provided information about the past climate. For example, the presence of broken, blacked, and burned clay pots in association with ash layers in an archeological site is an indicative of warm climate.

2.4.3 Geological Record

Geology is the branch of science that deals with the study of the Earth's formation, evolution, age, structure, composition, and processes that have shaped it over time (about 4.6 billion years ago). Geologists are specialists in the components of the Earth. They (geologists) have access to the Earth's material, which consists of various rock types (igneous, sedimentary, and metamorphic), fossils, sediments, and soils. This material, which is also known as rock record, provides a wealth of proxies and indirect data that can be used to recreate the history of the Earth's temperature throughout geological time (Evelick 2020). It should be remembered that the geological record predates historical and archeological evidence by a significant amount. Numerous short- and long-term climatic variations occurred during the 4.6 billion years of the Earth's history, leaving a variety of climatic proxies or natural archives recorded in the geological record. Sedimentary rock types, fossils, ice cores, and cave deposits are a few of the most significant natural archives.

2.4.3.1 Sedimentary Rock Types

The sedimentary rocks are formed by the gradual processes of sediment deposition brought by rivers and streams into oceans and other bodies of water (such as rivers or lakes), and over the course of millions of years, the soft sediment eventually got consolidated into stratified (layered) hard rocks (Evenick 2021). These rock bodies constitute the sedimentary rocks. Many climatically sensitive sedimentary rock types provide natural climatic archives as described below:

- *Glacial features*: Some glacial characteristics, such as striae, tillites, and moraines, are visible in the field and act as helpful climatic archives for the cold, glacial climate of high latitudes and elevations. As the glacier moves, it erodes/breaks rocks lying at its base and transports them in the direction of flow, leaving behind deep scratches in the underlying rocks, which are made by rock fragments carried by glaciers and are termed as glacier striae. As the glacier advances, it drops a mixture of sediments consisting of boulders, pebbles, sand, and mud, which later get settled by melt water of the glaciers; this heterogeneous mixture is known as till, and when lithified, it is known as tillites. As glaciers further advances, they form ridge-like deposits composed of unsorted mixture of fine rock particles to great boulders derived from the glacier which are known as moraines. The drumlins, kames, and eskers are other glacier features that also provide paleoclimatic information.
- *Rock types*: Some particular sedimentary rocks, like calcretes, which are accumulations of calcium carbonate, are formed by the near-surface evaporation of groundwater, and evaporites (Evenick 2021), which are composed of rock salt and are also known as halite and gypsum, are formed by the evaporation of surface water, which can help locate areas with arid, dry, and warm climates in mid to low latitudes (Evenick 2020). The sandstone formed by lithification of desert dunes, is characterized by large-scale cross-bedding tells us about desert likes condition and wind direction. The varves are lake deposits, consisting of alternating layers of coarse and fine-grained sediments that have deposited in lakes. The lake receives coarse-grained sediments at the time when sediment supply is high possibly in summer season and fine-grained when sediment supply is low in winter season. Therefore, alternating coarse- and fine-grained layering of sediments is considered to be associated with cyclic seasonal variation. The limestone (carbonate) rocks rich in coral remains indicate warm water (tropical ocean) having temperature ranging from 21 to 29 °C. The coal-bearing sedimentary rocks are indicative of humid tropical settings. Laterites, which are nodular soils that range in color from brown to red and are rich in iron, aluminum, and manganese, typically occur in hot, humid tropical climates with heavy rainfall.

2.4.3.2 Fossils

These are traces of ancient life that have been preserved in sedimentary rocks. It is well known that some species, notably animals and plants, are very reliant on their environment and that many of them have a limited capacity for adaptation to particular climatic circumstances. As a result, their fossils offer important hints about the past climate. Since reptiles (such as lizards and snakes) cannot survive in cold climates because their bodies are unable to maintain constant warm temperatures, their fossil remains are excellent markers of a warm climate. Due to the fact that modern cycads grow in tropical and subtropical climate zones, cycad fossils suggest that these regions were once home to tropical and subtropical climate. The edges of plant leaves are useful markers of past climate; for instance, smooth leaf margins on fossil leaves are good indicators of tropical environment, whereas leaf margins with toothed or lobed margins indicate cold climate.

The study of growth rings in trees and corals tells us about past seasonal variation. The trunk of a tree and skeleton of a sea coral contain numerous almost circular growth rings. In each season, a new ring adds and preserves weather conditions of the particular season. As they grow, many rings are added, which reflects season history of the area during the period of a tree or coral growth. Dendroclimatology is the study of growth rings in trees to determine climate. Coral clock is the study of the growth rings of coral. Coral growth rings provide evidence that the Earth's rotational speed has been gradually slowing down since ancient times because of the moon's gravitational pull.

2.4.3.3 Ice Cores

They include ice cores taken by drilling glaciers and ice sheets in the polar regions, northern Greenland, and high mountains of the Andes and the Himalaya that are persistently cold and experience little or no melting. By using a variety of techniques, the ice cores that were recovered by drilling are utilized to examine the air bubbles, water, and material trapped inside of them, including ancient atmospheric oxygen, hydrogen, and CO_2 as well as dust and ash particles. Ice cores are a good source of information about ancient climatic data (Gornitz 2009).

2.4.3.4 Cave Deposits

These speleothems (stalagmites, stalactites, and flowstones) are calcium carbonate deposits that have been formed in a limestone cave and could be an indicator of a non-glacial terrestrial climate. The speleothems are secondary mineral deposits formed from groundwater within underground caverns. The speleothems possesses different types of annual laminas and preserve the seasonality. The oxygen- and carbon-stable isotopic analyses of each lamina signify about the past rainfull, veg etation, and other climatic factors. Paleoclimatic data from cave deposits dates to about 30,000 years ago. Understanding natural archives and the techniques used in their investigation is necessary for reconstructing the past climate. Therefore, it is not necessary that our interpretation of proxy data is always accurate. Knowing how different geological climate proxies relate to the current climate can help us get over the challenges each of the aforementioned climate proxies has when attempting to infer historical climatic information.

2.5 Climate of the Quaternary Period

Pleistocene and Holocene are two of the epochs that make up the Quaternary, the youngest phase of the Cenozoic era (Table 2.1). It is a period of greater climate changes since the past 60 million years (Bradley 2015). The climate on Earth has changed dramatically over the past 2.58 million years, notably in the Northern Hemisphere, parts of Antarctica, and high mountainous regions, which have repeatedly witnessed widespread glaciations. As a result, this time period is known as "the Great Ice Age." In reality, we are still in the interglacial (warm) period of this epoch, which means it has not yet ended. Climate proxies are often used to recreate the climate of the Quaternary Period. These include tree growth rings, cave and glacial features, lakes and dune deposits, microfossils, pollen grains, and ice cores.

2.5.1 Pleistocene

The Pleistocene epoch spans the period 2.58 million to 11,700 years ago. Epoch is frequently used to refer to a division or unit of time on the geological time scale. The analysis of numerous Pleistocene climatic proxies makes it abundantly evident that both the advent of humanity and significant climate shifts occurred during this time. Around 2.5 million years ago, there was a change in the climate on a global scale, resulting in a cooler climate. In response to this climate change, the genus *Homo* (i.e., humans) developed from australopithecine (ape- and human-like primates) ancestors. Additionally, the appearance of the mammalian genera *Bos* (bovid), *Elephas* (elephant), and *Equus* (horse) marks the beginning of the Pleistocene. The climate of the Pleistocene is characterized by an orderly sequence of interglacial-glacial-interglacial periods. During this epoch, the cold climate intensified, which led to the development of extensive ice sheets and mountain glaciers in high latitude and high altitude regions of the Earth. It includes a larger part of the Northern Hemisphere (the United States, Canada, Greenland, Europe, Asia, and northern Russia), Antarctica, South America, and mountainous areas of the Rockies, Alps, Himalaya, Kilimanjaro, and Mount Kenya. The maximum Pleistocene glaciations occurred in the Northern Hemisphere. Therefore, the massive ice sheets covering the parts of Eastern North America, Western North America, and Northern Europe are termed as the Laurentide ice sheet, Cordilleran ice sheet, and Scandinavian ice sheet, respectively. It's interesting to note that during the Pleistocene period, glaciers and ice sheets covered around 30% of the planet's surface, and there have been at least 20 known cycles of glacial and inter-glacial stages. Glacial stage refers to the cold period when glaciers are highly extensive, while interglacial stage refers to the warm, dry period between two intervening glacial stages when glaciers are less extensive.

The tropical regions endured the heaviest rainfall and the most humid climate throughout the Pleistocene. The period of maximum rainfall known as pluvial period and intervening period of dry climate between two successive pluvial periods is described as inter-pluvial period. The pluvial periods as stated above experienced maximum rainfall, caused the flooding in the rivers and streams, and formed extensive flood plain deposits consisting of layers of sand, silt, and gravel. It is observed that glacial-interglacial and pluvial-inter-pluvial intervals are interrelated. The deciduous and coniferous forests were more common during the warm period; however, grasses, lichens, and mosses dominated land during the winter period. Geologically, the Pleistocene epoch has been classified into three subdivisions: lower, middle, and upper. And, each of these subdivisions had experienced episodes of glacial-interglacial. In the Pleistocene, there were four glacial and three interglacial periods (Table 2.2).

The four pluvial periods such as Kageran, Kamasian, Kanjeran, and Gamblian have been recorded from Africa, and their occurrence was corresponding to the occurrences of Gunz, Mindel, Riss, and Wurm European glacial stages. It is believed that many mammalian faunas such as woolly rhinoceros, woolly mammoth, Columbian mammoth, cave lion, and rein deer adapted to the cold climatic conditions.

Table 2.2 Glacial-interglacial periods of the Pleistocene

Epoch		Glacial-interglacial stage	
		Europe	North America
	Holocene	Interglacial	Interglacial
Pleistocene	Upper (12,600 to 11,700 years ago)	Wurm glaciation	Wisconsinan glaciation
		Riss-Wurm interglacial	Sangamonian interglacial
		Riss glaciation	Illinoian glaciation
	Middle (78,100 to 12,600 years ago)	Mindel-Riss interglacial	Yarmouthian interglacial
		Mindel glaciation	Kansan glaciation
		Gunz-Mindel interglacial	Aftonian interglacial
	Lower (2.58 million to 78,100 years ago)	Gunz glaciation	Nebraskan glaciation
Pliocene	–	–	–

2.5.2   *Holocene*

The most recent period on the geologic time scale is called the Holocene which is still continuing (Table 2.1). It begins roughly 11,700 years ago, with the conclusion of the last major glacial stage of the Pleistocene, and continues to the present day. It is separated into three age groups (Table 2.3). It is relatively a warm period during which human influences had been significantly altered the Earth system particularly its environment. Initially, humans altered the Earth's environment by hunting, cutting down trees, farming (agriculture), latterly establishing civilization, building towns and cities, industries with burning of fossil fuels, extracting natural resources, and finally establishing huge networks of transportation and communication systems (Stanley 2009). It is noted that humanity has broadly influenced the Holocene environment of the Earth; therefore, it is sometimes also known as Anthropocene. The term Anthropocene is an informal name, and till date Holocene is a valid epoch.

The Anthropocene refers to "Age of Man." In simple words, the Anthropocene can be described as the geology of humanity, which focuses on the cumulative role of humans as geologic and geomorphic agents in altering the Earth's environment by multiple ways such as through agriculture, mining, industrialization, urbanization, or globalization. The word Anthropocene was broadly used in the scientific literature of China during the 1990s in an informal way. In 2000, Paul Crutzen and Eugene Stoermer formally presented Anthropocene and also discussed it in the context of geological time scale. The Anthropocene is a less popular concept as compared to global warming (Syvitski 2012). The Anthropocene is still an informal time unit, and its beginning is still a matter of debate, but many workers believe that it began with the Industrial Revolution in Europe around 1800 years before present (Zalasiewicz et al. 2009).

The Holocene epoch is very important for us because it shows how the Earth's environment reached to its present form. It also experienced varied cycles of climate change (Table 2.3). It should be emphasized that the radiocarbon dating method, which uses carbon-14 and has a half-life of 5730 years, is a highly accurate way to date Holocene sediments and organic remains. The Early Holocene (11,700 to 8200 years before present) was a time of global warming and moist conditions prevailed in tropical dessert areas. About three episodes of above-sea-level elevation were recorded during this interval based on remains of reef-building sea corals. The dry interval of the Early Holocene is described as Boreal period and wet as Atlantic period (Table 2.4).

Table 2.3   Holocene time scale

Period	Epoch	Age	Duration in years
Quaternary	Holocene	Meghalayan (late)	4200 to present
		Northgrippian (middle)	8200 to 4200
		Greenlandian (early)	11,700 to 8200
	Pleistocene	Upper	12,600 to 11,700

Table 2.4 Holocene climate (modified after Mathur 2005)

Epoch	Glacial stage	Environmental period	Age-based carbon-14 method (in years before present)	Climate
Holocene	Post glacial	Subatlantic	2500 to 0	Wet and cool
		Subboreal	5000 to 2500	Dry and warm
		Atlantic	8000 to 5000	Wet and warm
		Boreal	10,000 to 8000	Dry and warm
Pleistocene	Late glacial	–		

The Middle Holocene was a time of high warming, and global temperature rose by 4 °C to 5 °C. During this interval, Arabia and India experienced higher monsoon circulation (Mathur 2005). During the Early and Middle Holocene between 9000 and 6000 years before present, many continental glaciers disappeared. The dry and warm climate of the Middle Holocene is termed as Subboreal environmental period (Table 2.4). Rapid periods of warming and cooling occurred during the Late Holocene (4200 years ago to the present). Between 1445 and 1700 AD, the Arctic region was covered by ice, and many glaciers advanced which gave rise to Little Ice Age. This evidence indicates that the climate has been changing in the Late Holocene. The Subatlantic environmental period is named after the cool, wet climate of the Late Holocene.

In a nutshell, we live in the Holocene. This epoch possesses relatively high sea level, minimal ice covers (which are still extensive in polar regions and high elevation of the mountainous regions), mid-latitude deciduous forest, and huge expansion of human population (Bloom 2009). The modern and industrial societies of humans have continuously been altering the Earth's environment by burning fossil fuels and adding high concentration of CO_2 as a byproduct of fossil fuel combustion into the atmosphere. It is altering the climate system and, thus, causing the global warming.

References

Barry RG, Chorley RI (2010) Atmosphere, weather and climate. Routledge, New York

Beerling DJ, Berner RA (2005) Feedbacks and the coevolution of plants and atmospheric CO_2. Proc Natl Acad Sci U S A 102:302–1305

Bloom AL (2009) Geomorphology-a systematic analysis of late Cenozoic landforms. Phi Learning Private Ltd., New Delhi

Bradley RS (2015) Paleoclimatology - reconstructing climates of the quaternary. Academic Press, Amsterdam

Evenick JC (2020) Late cretaceous (Cenomanian and Turonian) organofacies and TOC maps: example of leveraging the global rise in public-domain geochemical source rock data. Mar Pet Geol 111:301–308, ISSN 0264-8172

Evenick JC (2021) Glimpses into Earth's history using a revised global sedimentary basin map. Earth Sci Rev 215:103564, ISSN 0012-8252

Gornitz V (2009) Paleoclimate proxies, an introduction. In: Gornitz V (ed) Encyclopedia of paleoclimatology and ancient environments. Encyclopedia of earth sciences series. Springer, Dordrecht

Krebs RE (2007) The basics of earth science. Geenwood Press, Westport

Mathur UB (2005) Quaternary geology - Indian perspective. Geological Spective. Geological Society of India, Bangalore

Ozturk M, Hakeem KR, Faridah-Hanum I, Efe R (2015) Climate change impacts on high-altitude ecosystems. Springer Science+Business Media, XVII, New York, 695 pp

Reitz EJ, Newsom LA, Scudder SJ, Scarry CM (2008) Introduction to environmental archaeology. In: Case studies in environmental archaeology. Springer, pp 3–19

Ruddiman WF (2001) Earth's climate-past and future. WH Freeman and Company, New York

Stanley SM (2009) Earth system history WII. Freeman and Company, New York

Syvitski J (2012) Anthropocene; an epoch of our making. Glob Change 78:12–15

Zalasiewicz J, Waters CN, Summerhayes CP (2009) The anthropocene as a geological time unit. Cambridge University Press, Cambridge. https://www.ipcc.ch/site/assets/uploads/2018/02/WGILAR5-AnnexII_FINAL.pdf

Chapter 3
Sources and Monitoring Tools of Atmospheric Carbon Dioxide

3.1 Introduction

Carbon dioxide (CO_2), methane (CH_4), nitrous oxide (N_2O), and water vapor are examples of the greenhouse gases that trap solar energy and keep the Earth habitable. However, a few synthetic materials also exhibit this ability to absorb infrared wavelengths, such as sulfur hexafluoride (SF_6), perfluorocarbons (PFCs), hydrofluorocarbons (HFCs), and chlorofluorocarbons (CFCs). The global surface temperature might fall below −18 °C if there are no greenhouse gases (GHGs) in the atmosphere of our planet (Casper 2010). Thus, any variations in the concentrations of the listed GHGs would eventually have an effect on the Earth's surface temperature. Additionally, the GHGs in the atmosphere possess different lifetimes and capacities for absorbing heat from terrestrial radiation (Casper 2010). As a result, the global warming potential (GWP) of the GHGs varies depending on how long they had been present in the atmosphere. According to the US Environmental Protection Agency (EPA), the HFCs, PFCs, and SF_6 gases are more efficient at trapping infrared wavelengths and have a longer residence time. The various sources of GHGs and monitoring techniques to estimate CO_2 emissions include infrared analyzers, manometry, and satellite interventions which are discussed in this chapter.

3.2 Sources of GHGs

The majority of GHGs, like CO_2, are found in nature and cycle through processes in the global biogeochemical system (regional to global scale) which have evolved as a result of the Industrial Revolution and intensive agriculture practices (Lal 2004a). The most significant GHGs in the atmosphere are CO_2, CH_4, and N_2O, and human activity has greatly raised their concentrations in the atmosphere (IPCC 2001).

© The Author(s), under exclusive license to Springer Nature Switzerland AG 2023
M. A. Dervash et al., *Phytosequestration*, SpringerBriefs in Environmental
Science, https://doi.org/10.1007/978-3-031-26921-9_3

3.2.1   Carbon Dioxide (CO₂)

Many scientists in the nineteenth century believed that atmospheric CO_2 causes the greenhouse effect and regulates or affects the Earth's surface temperature (IPCC 2001). The CO_2 naturally escapes into the atmosphere as a result of burning fossil fuels, mining coal, forest fires, alterations in land use and land cover, the breakdown of soil organic matter, and plant respiration (Lal 2008). The amount of CO_2 in the atmosphere has escalated as a result of human activities, rising from 280 ppm (pre-industrial level) to 410 ppm (NOAA 2019). The burning of fossil fuels and altering land use release annual concentration of around 11.3 ± 0.9 GtC (41.5 ± 3 GtCO$_2$) into the atmosphere (Le Querre et al. 2018). CO_2 levels in the atmosphere are increasing at an alarming rate of 1.5 ppm (Lal 2004b).

3.2.2   Methane (CH₄)

CH_4 is produced when plant debris breaks down anaerobically (without oxygen) during the decomposition process. The GWP for CH_4 is 23 times greater than that of CO_2 (Casper 2010). CH_4 emissions from livestock, including sheep, goats, cows, buffalo, camels, and bison, are also substantial contributors to the atmosphere. The study period, between 2003 and 2012, is predicted to have resulted in 558 Tg CH_4 in annual worldwide CH_4 emissions (Saunois et al. 2016). Around 60% of the world's CH_4 emissions are caused by humans. Livestock and agricultural activities provide between 70 and 90% of the world's CH_4 emissions (Gerber et al. 2013). Permafrost, wetlands, freshwater bodies, oceans, soils, flooded paddy fields, and wildfires account for around 10% of total CH_4 emissions (Kirschke et al. 2013). India accounts for 15.3 TgYr^{-1} or roughly 2.74% of the world's annual CH_4 emissions mainly from livestock (Kumari et al. 2018). More crucially, CH_4 emissions are rising annually at rates of 0.90 kg for the world and 1.10 kg for India, respectively (Patra 2014). In this perspective, it is predicted that the CH_4 emission rate may be 105 kg and 120 kg, respectively, in the years 2025 and 2050 (Patra 2014). Naturally, areas and nations with larger herds of cattle produce more CH_4 than other areas and nations (Knapp et al. 2014). According to Tian et al. (2015), the rapid rise in CH_4 emissions from paddy fields and natural wetlands is also a result of global warming.

3.2.3   Nitrous Oxide (N₂O)

Parts per billion by volume (ppbv) of N_2O in the atmosphere increased from 288 to 322.5 ppbv (Lal 2004a, b; Butterbach-Bahl et al. 2013). Due to human activity, it is rising in the atmosphere at a rate of 0.77 ppbv every year (Butterbach-Bahl et al. 2013). The use of nitrogen-based synthetic fertilizers and microbiological processes in soils, sediments, water bodies, animal manures, and other environments are the principal sources of N_2O (Tian et al. 2015). Animal manure management and the use of nitrogen fertilizers in agriculture each generate about 4.3–5.8 Tg N_2O-N yr^{-1} of emissions into the atmosphere (Syakila and Kroeze 2011). The atmosphere receives 6–7 Tg N_2O-N $yr.^{-1}$ emissions from natural soils (Syakila and Kroeze 2011). Additionally, agriculture and livestock operations are responsible for nearly two-thirds of all anthropogenic N_2O emissions into the atmosphere (Steinfeld et al. 2006).

3.2.4   Chlorofluorocarbons (CFCs), Hydrofluorocarbons (HFCs), and Perfluorocarbons (PFCs)

The CFC is a synthetic compound made up of fluorine, chlorine, and carbon. A substance is referred to be HFCs if it contains hydrogen. They are frequently utilized as refrigerants, cooling agents, aerosol propellants for medications, cleaning agents, solvent, and degreasing solutions, as well as blowing agents for foams. The stratospheric ozone layers are significantly degraded by the CFCs. As it is an established fact, living creatures are safeguarded from the sun's harmful ultraviolet (UV) radiation by the stratospheric ozone layer. The chlorine radicals are released into the atmosphere when the CFC molecules are broken down by solar radiation. Consequently, the chlorine radicals react with the ozone molecules, destroying the stratospheric ozone layer (one chlorine radical destroys one lakh ozone molecules). The CFCs possess long residence time in the atmosphere. To control the industry's usage of CFCs for ozone layer protection, the Montreal Protocol was established in 1987. The refrigerant industries began utilizing HFCs to replace CFCs. The HFCs do not harm the ozone layer since fluorine is used in place of chlorine, but they possess GWP 10,000 times greater than CO_2 (Casper 2010). Perfluorocarbons (PFCs) are currently employed as solvents, water repellents, anesthetics in surgery, coolants in freezers, and chemical reagents (Casper 2010).

3.2.5   Sulphur Hexafluoride (SF₆)

One of the most significant GHGs is the SF_6, which has a GWP 23,900-fold greater than CO_2. The primary producers of SF_6 emissions are the electrical equipment and magnesium smelting industries (Zhou et al. 2018). In comparison to anthropogenic sources, the natural supply of SF_6 is practically minimal (Vollmer et al. 2002). The SF_6 contributes 0.1% of all anthropogenic radioactive forcing, and it is constantly escalating in the atmosphere (Rigby et al. 2010). Around 1.25 Gt of CO_2 was predicted to be the total amount of SF_6 emissions worldwide in 2010 (UNFCCC). The electrical sectors of developing countries account for the majority of SF_6 emissions, which are anticipated to rise in the upcoming years (Fang et al. 2013). In order to create a mitigation strategy, it is therefore even more important to comprehend the SF_6 emission of developing nations.

3.3 Greenhouse Gas Emission Inventories

Since 1996, the "Annex I (developed nations)" and "Annex II (developing countries)" parties to the United Nations Framework Convention on Climate Change (UNFCCC) have published the periodic national inventories of greenhouse gas emissions and removals (Jonas et al. 2010). Additionally, in order for a country to recognize and enumerate the sources and sinks of GHGs in an attempt to address climate change, GHG emission inventories are important (IPCC 2006). While policymakers use this inventory to design measures to lower GHG emissions, scientists use it to create accurate atmospheric models. As per the Intergovernmental Panel on Climate Change (IPCC) (2006), the recommendations for greenhouse gas inventories have been updated to include nitrogen trifluoride (NF_3), SF_6, CO_2, CH_4, N_2O, HFCs, and PFCs, trifluoromethyl sulfur pentafluoride (SF_5CF_3), halogenated ethers, and halogenated carbons. According to Sharma et al. (2006), the emission data from the ratified countries aids in determining the contribution of those countries' GHG emissions to global GHG emissions as well as the development of plans and policies for reducing those emissions.

Five separate sectors with a variety of categories are used to estimate the GHG emission inventories. Energy; Industrial Processes and Product Use (IPPU); Agriculture, Forestry, and Other Land Use (AFOLU); waste; and others are listed in that order (IPCC 2006). Therefore, a country would create an emission inventory using the IPCC standards, going from a lower level (categories) to a higher level (categories) (IPCC 2006). In order to include the energy, agricultural, and waste sectors with subcategories, the first complete IPCC greenhouse gas emission inventory guidelines were upgraded with Good Practice Guidelines (GPG) in 2000 (Sharma et al. 2011). The updated IPCC recommendations for a greenhouse gas inventory were released in 2006. There are numerous subcategories in each area. For instance, in the energy sector, the category of transportation includes two-wheelers as a subcategory. Thus, a country may eliminate the data uncertainty in each sector.

3.4 Greenhouse Gas Emission Estimation Methods

Based on the amount of data needed and the complexity of the analysis, the IPCC has created three varied methodological strategies (tier 1, tier 2, and tier 3) (IPCC 2006). The tier 1 technique makes use of rough data from global land use and land cover maps, deforestation rates, agricultural production data from statistics, and other IPCC criteria. The tier 2 approach, however, uses parameters that are more country-specific, such as emission factors, for instance, data on emissions from deforestation and changes in land use and land cover in various climatic zones of the nation. The tier 3 approach employs higher-order techniques, including models and information from measuring systems for the national forest inventory that are adapted to take into account local conditions (Sharma et al. 2011). The tier methodologies listed above are typically used to lessen the vagueness of GHG estimations from various sectors of a country. Sharma et al. (2011) provided information on India's GHG emissions from various anthropogenic sources and sinks.

3.5   Greenhouse Gas Monitoring

It is the process of directly measuring greenhouse gas emissions and levels. The atmospheric CO_2 can be measured quantitatively using a variety of techniques, such as infrared analysis and manometry. Other equipments are used to measure CH_4 and N_2O. The Orbiting Carbon Observatory and ground station networks like the Integrated Carbon Observation System (ICOS) use space-based instruments to study greenhouse gases.

In order to precisely estimate the methane emissions from different sources, differential absorption lidar (DIAL) conducts vertical scans above CH_4 sources and then spatially isolates the scans. CH_4 is one of the most influential gaseous hydrocarbon species; hence measuring methane emissions is essential for climate change studies.

N_2O concentrations in the upper to lower troposphere are estimated by using the Atmospheric Chemistry Experiment-Fourier Transform Spectrometer (ACE-FTS). This experiment, which is connected to the Canadian satellite SCISAT, has demonstrated that N_2O is present year-round throughout the whole atmosphere, mostly as a result of energetic particle precipitation. The instrument's measurements reveal that N_2O is produced in the lower thermosphere by different reactions than it is in the mid- to upper mesosphere. By analyzing the many ways that N_2O is released into the atmosphere, the ACE-FTS is an essential tool for forecasting future ozone depletion in the high stratosphere.

3.6   CO₂ Monitoring

Accurate monitoring of atmospheric CO_2 and its spatiotemporal change is necessary to ascertain the distribution and dynamics of carbon sources and sinks at regional and global scales (Mustafa et al. 2021). The way that CO_2 (and other greenhouse gases) that contains two or more atoms of different elements absorbs IR light is unique and distinct. Between 1958 and 2006, infrared analyzers were employed at Mauna Loa Observatory and Scripps Institution of Oceanography. IR analyzers operate by pumping an unidentified sample of dry air through a 40-cm-long chamber. Air devoid of CO_2 is present in a reference cell. Broadband infrared radiation from a glowing nichrome filament divides into two beams and goes through the gas cells. Due to the radiation's partial absorption by CO_2, radiation from the reference cell is able to reach the detector at a higher rate than radiation from the sample cell. A strip chart recorder is used to collect the data. By calibrating with a reference gas that has a known amount of CO_2, the concentration of CO_2 in the sample is estimated. IR sensors can monitor gases including CO_2, CH_4, and water vapor. As a result, IR detectors are most frequently utilized in CO_2 analyzers. Because the measured gas doesn't establish contact with the sensor and because they are stable and highly selective for CO_2, they can withstand adverse environments including high humidity, dust, and filth. Their lifespan is likewise very long. Moreover, accuracy of instruments used to monitor atmospheric carbon dioxide concentrations must be at least 1 ppmv.

Several ground-based stations, including those in the Global Atmospheric Watch (GAW) network and the Total Carbon Column Observing Network (TCCON) locations, are keeping close tabs on atmospheric CO_2 levels (Mendonca et al. 2019). However, these observational sites are insufficient to precisely monitor atmospheric CO_2 at local and global scales due to their uneven spatial distribution and low spatial coverage (Kulawik et al. 2016). The first two CO_2 monitoring satellites that were successfully launched into orbit were Greenhouse gases Observing SATellite (GOSAT) and the Orbiting Carbon Observatory-2 (OCO-2). Both instruments measure the optical depth of CO_2 using bands with centers at 1.6 and 2.0 m, respectively, and O_2 using band A with a center at 0.76 m (Kiel et al. 2019). Another useful instrument for tracking atmospheric CO_2 and other variables is the integrated path differential absorption (IPDA) light detection and ranging (lidar) system (Zhu et al. 2020).

The Integrated Carbon Observation System (ICOS) was founded as a European Research Infrastructure Consortium (ERIC) in Helsinki, Finland, in October 2015. The primary objective of ICOS is to create an Integrated Carbon Observation System Research Infrastructure (ICOSRI), which makes it easier to conduct studies on greenhouse gas emissions, sinks, as well as sources. To develop logical data products and to advance learning and innovation, the ICOS ERIC works to connect its own exploration with those of other researchers studying greenhouse gas emissions.

In order to quantify atmospheric CO_2, manometry first determines the quantity, temperature, and pressure of a specific volume of dry air which is then dried by being forced through many dry ice traps before being collected in a container of 5-liter capacity. A thermometer is used to measure the temperature, and manometry is used to determine the pressure (Harris 2010). The CO_2 then begins to condense and become volume quantifiable after the addition of liquid nitrogen. With this pressure, the ideal gas law is accurate to 0.3% (Harris 2010). Another technique for monitoring atmospheric CO_2 is titration, which was originally applied at 15 separate ground stations by a Scandinavian team. They started by putting a 100.0-mL air sample through a barium hydroxide solution that also contained cresolphthalein indicator (Harris 2010).

References

Butterbach-Bahl K, Baggs EM, Dannenmann M, Kiese R, Zechmeister-Boltenstern S (2013) Nitrous oxide emissions from soils: how well do we understand the processes and their controls? Philos Trans R Soc B 368:20130122

Casper JK (2010) Greenhouse gases: worldwide impacts (global warming). Facts on File Inc, New York

Fang X, Thompson RL, Saito T, Yokouchi Y, Kim J, Li S, Kim KR, Park S, Graziosi F, Stohl A (2013) Sulfur hexafluoride (SF6) emissions in East Asia determined by inverse modeling. Atmos Chem Phys Discuss 13:21003–21040

Gerber PJ, Hristov AN, Henderson B, Makkar H, Oh J, Lee C, Meinen R, Montes F, Ott T, Firkins J, Rotz A, Dell C, Adesogan AT, Yang WZ, Tricarico JM, Kebreab E, Waghorn G, Dijkstra J, Oosting S (2013) Technical options for the mitigation of direct methane and nitrous oxide emissions from livestock: a review. Animal 7:220–234

Harris DC (2010) Charles David Keeling and the story of atmospheric CO_2 measurements. Anal Chem 82:7865–7870

IPCC (2001) Climate change 2001: the scientific basis. In: Houghton JT, Ding Y, Griggs DJ, Noguer M, van der Linden PJ, Dai X, Maskell K, Johnson CA (eds) Contribution of working group I to the third assessment report of the intergovernmental panel on climate change. Cambridge University Press, Cambridge/New York, p 88

IPCC (2006) IPCC guidelines for National Greenhouse Gas Inventories- a primer, Prepared by the National Greenhouse Gas Inventories Programme, Eggleston HS, Miwa K, Srivastava N, Tanabe K (eds). Published: IGES

Jonas M, Bun R, Nahorski Z, Marland G, Gusti M, Danylo O (2010) Quantifying greenhouse gas emissions. Mitig Adapt Strateg Glob Chang 24:839–852

Kiel M, O'Dell CW, Fisher B, Eldering A, Nassar R, MacDonald CG, Wennberg PO (2019) How bias correction goes wrong: measurement of XCO$_2$ affected by erroneous surface pressure estimates. Atmos Meas Tech 12.2241–2259

Kirschke S, Bousquet P, Ciais P, Saunois M, Canadell JG et al (2013) Three decades of global methane sources and sinks. Nat Geosci 6:813–823

Knapp JR, Laur GL, Vadas PA, Weiss WP, Tricarico JM (2014) Enteric methane in dairy cattle production: quantifying the opportunities and impact of reducing emissions. J Dairy Sci 97:3231–3261

Kulawik S, Wunch D, O'Dell C, Frakenberg C, Reuter M, Oda T et al (2016) Consistent evaluation of ACOS-GOSAT, BESD-SCIAMACHY, CarbonTracker, and MACC through comparisons to TCCON. Atmos Meas Tech 9:683–709

Kumari P, Hiloidhari M, Naik SN, Dahiya RP (2018) Methane emission assessment from Indian livestock and its role in climate change using climate metrics. In: Hussain S (ed) Climate change and agriculture. Intech Open, p 248

Lal R (2004a) Agricultural activities and the global carbon cycle. Nutr Cycl Agroecosys 70:103–116

Lal R (2004b) Soil carbon sequestration to mitigate climate change. Geoderma 123:1–22

Lal R (2008) Carbon sequestration. Philos Trans R Soc B 363:815–830

Le Querre C, Andrew RM, Friedlingstein P, Sitch S, Hauck J, Pongratz J, Pickers PA, Korsbakken JI, Peters GP, Canadell JG, Arneth A, Arora VK, Barbero L, Bastos A, Bopp L, Chevallier F, Chini LP, Ciais P, Doney SC, Gkritzalis T, Goll DS, Harris I, Haverd V, Hoffman FM, Hoppema M, Houghton RA, Hurtt G, Ilyina T, Jain AK, Johannessen T, Jones CD, Kato E, Keeling RF, Goldewijk KK, Landschützer P, Lefèvre N, Lienert S, Liu Z, Lombardozzi D, Metzl N, Munro DR, Nabel JEMS, Nakaoka S, Neill C, Olsen A, Ono T, Patra P, Peregon A, Peters W, Peylin P, Pfeil B, Pierrot D, Poulter B, Rehder G, Resplandy L, Robertson E, Rocher M, Rödenbeck C, Schuster U, Schwinger J, Séférian R, Skjelvan I, Steinhoff T, Sutton A, Tans PP, Tian H, Tilbrook B, Tubiello FN, van der Laan-Luijkx IT, van der Werf GR, Viovy N, Walker AP, Wiltshire AJ, Wright R, Zaehle S, Zheng B (2018) Global carbon budget. Earth Syst Sci Data 10:2141–2194

Mendonca J, Strong K, Wunch D, Toon GC, Long DA, Hodges JT, Sironneau VT, Franklin JE (2019) Using a speed-dependent Voigt line shape to retrieve O_2 from Total carbon column observing network solar spectra to improve measurements of XCO_2. Atmos Meas Tech 12:10.5194

Mustafa F, Bu L, Wang Q, Yao N, Shahzaman M, Bilal M, Aslam RW, Iqbal R (2021) Neural-network-based estimation of regional-scale anthropogenic CO_2 emissions using an Orbiting Carbon Observatory-2 (OCO-2) dataset over East and West Asia. Atmos Meas Tech 14:7277–7290

NOAA (National Oceanic and Atmospheric Administration) (2019) Atmospheric CO_2 Mauna Loa Observatory (Scripps / NOAA / ESRL). Monthly & Annual Mean CO_2 Concentrations (ppm). Washington, DC

Patra AK (2014) Trends and projected estimates of GHG emissions from Indian livestock in comparisons with GHG emissions from world and developing countries. Asian-Australas J Anim Sci 27:592–599

Rigby M, Muhle J, Miller BR, Prinn RG, Krummel BR et al (2010) History of atmospheric SF6 from 1973 to 2008. Atmos Chem Phys 10:10305–10320

Saunois M, Bousquet P, Poulter B, Peregon A, Ciais P et al (2016) The global methane budget 2000–2012. Earth Syst Sci Data 8:697–751

Sharma S, Bhattacharya S, Garg A (2006) Greenhouse gas emissions from India: a perspective. Curr Sci 90:326–333

Sharma SK, Choudhary A, Sarkar P, Biswas S, Singh A et al (2011) Greenhouse gas inventory estimates for India. Curr Sci 101:405–415

Steinfeld H, Gerber P, Wassenaar T, Castel V, Rosales M, de Haan C (2006) Livestock's long shadow. FAO, Rome

Syakila A, Kroeze C (2011) The global nitrous oxide budget revisited. Greenh Gas Meas Manag 1:17–26

Tian M, Ajay VS, Dunzhu D, Hameed SS, Li X, Liu Z et al (2015) A cluster-randomized controlled trial of a simplified multifaceted management program for individuals at high cardiovascular risk (simcard trial) in rural Tibet, China and Haryana. India 132:815–824

Vollmer M, Weiss RF, Schlosser P, Willams RT (2002) Deep-water renewal in Lake Issyk-Kul. Geophys Res Lett 29:1–4

Zhou S, Teng F, Tong Q (2018) Mitigating Sulfur Hexafluoride (SF6) emission from electrical equipment in China. Sustainability 10:2402

Zhu Y, Yang J, Chen X, Zhu X, Zhang J, Li S, Sun Y, Hou X, Bi D, Bu L, Zhang Y, Liu J, Chen W (2020) Airborne validation experiment of 1.57-μm double-pulse IPDA LIDAR for atmospheric carbon dioxide measurement. Remote Sens 12:1999

Chapter 4
Global Warming: Impacts of Temperature Escalation

4.1 Introduction

Since the Industrial Revolution, anthropogenic activities like industrial, economic, land use, and population growth have significantly impacted climate by increasing atmospheric CO_2 and other heat-trapping gases. In 2010, CO_2 continued to be the main anthropogenic GHG, contributing 76% (383.8 Gt CO_2 eq/year) of all anthropogenic GHG emissions. CH_4 contributes 16% (781.6 Gt CO_2 eq/year), N_2O contributes 6.2% (311.9 Gt CO_2 eq/year), and fluorinated gases contribute 2.0% (100.2Gt CO_2 eq/year). According to the IPCC, by 2100, the average surface temperature could have increased by more than 6 °C. From 280 parts per million in the pre-industrial era to 400 parts per million in 2014, the atmospheric CO_2 concentration grew by 42.8% over the course of the twentieth century, and an increase in CO_2 concentration and other GHGs (CH_4, N_2O) led to an increase in the global mean temperature of 0.85 °C (0.65–1.06 °C) (IPCC 2014) (Fig. 4.1).

Continued GHG emissions will result in additional warming of the atmosphere, changes to various climate system components, significant livelihood consequences, and widespread and irreversible effects on people as well as terrestrial and aquatic ecosystem. For natural and human systems, climate change will both increase already present dangers and generate new ones (Ozturk et al. 2011). Risks are unevenly distributed and typically have a greater impact on underprivileged individuals and communities in nations of all developmental stages. Extreme weather events and unexpected precipitation will influence the effects of climate change and have a negative impact on soil, plants, people, and animals. Additionally, it is believed that climate change will increase the severity and frequency of floods, wildfires, glacier melt, dry spells, and the introduction of new diseases, crop shifting, starvation, and insect-pest infestations (Lal 2011). This chapter will provide an information regarding the impacts of climate change on terrestrial, as well as aquatic ecosystems.

© The Author(s), under exclusive license to Springer Nature Switzerland AG 2023
M. A. Dervash et al., *Phytosequestration*, SpringerBriefs in Environmental
Science, https://doi.org/10.1007/978-3-031-26921-9_4

Fig. 4.1 Impacts of global warming. (Source: National Oceanic and Atmospheric Administration (NOAA) https://images.app.goo.gl/kPNM2TsLwAnqbMgw7)

4.2   Climate Change Impacts on Terrestrial Ecosystems

4.2.1   Carbon and Nitrogen Dynamics

The physical, chemical, and biological characteristics of soil offer details about factors affecting germination, root growth, and erosion processes, as well as information about water, air, temperature, microbial activity, and soil reactions. Other chemical and biological activities are supported by a number of soil physical qualities, which may then be further influenced by climate, landscape position, and land use. Significant soil indicators of climate change include mineralization, volatilization, microbial decomposition, salinization, evapotranspiration, and increased greenhouse gas emissions. The carbon and nitrogen dynamics of soil are highlighted as potential soil health determinants.

In climate change settings, the amount of organic matter in the soil is a sign of the quality of the organic matter and the health of the soil, acting as a nutritional agent for microbes during the nutrient cycle process (Gregorich et al. 1994). Reduced SOM can result in microbial biodiversity, duct ion infertility, loss of soil structure, decreased WHC, and an increase in soil erosion. SOM is what controls how the soil operates (Weil and Magdoff 2004). Farm-level land use changes and soil management techniques that encourage the accumulation of organic matter in soil will aid in absorbing CO_2 from the atmosphere, hence reducing global warming and the anomalies associated with climate change. Organic matter can play a crucial part in reducing the effects of flooding during high rainfall events by enhancing soil moisture and water storage. It can also conserve water during droughts, enhancing soil resilience.

4.2.2 Impacts of Climate Change on Soil Salinization

Although salt is required for cooking and food preparation, too much salt in the soil can kill crops and make fields unusable. In both rainfed and irrigated settings, soil salinization is a degrading process that destroys both crop productivity and soil fertility. Higher and more variable temperatures, precipitation patterns, and a greater frequency of extreme occurrences are the main predicted effects of climate change (Al-Najar and Ashour 2012). Water security will be threatened by declining water supplies connected to salt intrusion into surface and subterranean water. Therefore, controlling sustainable water resource consumption depends on recognizing the connection between surface and subterranean water. Food security will be significantly impacted by worldwide soil salinization and water constraint (Teh and Koh 2016). The effects of climate change and global warming are perceived worldwide, but they are more critical in arid regions where soil salinization is most likely to occur. The soils of Central Asia greatly speed up the formation of salt. The manner that salt is redistributed in the soil profile is determined by climatic factors, particularly precipitation. Rainfall in the winter and spring means that readily soluble salts are carried across more quickly. The following factors will cause climatic changes to widen the areas with saline soils: first, rising temperatures and aridity have a direct impact on salt transport and soil salt balance; second, changes in land use have an indirect impact on salt dynamics (Szabolcs 1990). Low-lying coastal areas commonly experience salt water inundation as a result of sea level rise, which continuously contaminates the soil nearby. Rainfall can disperse these salts, but climate change is also increasing the frequency and intensity of extreme weather events, such as heat waves and droughts. Due to increased groundwater consumption for irrigation and drinking, the water table is further drained, increasing the amount of salt that can seep into the soil. The warming of the water and rising ocean temperatures from climate change have an impact on soil salinity. Even with greater reductions in GHG emissions, scientists now forecast that global mean sea levels would rise by at least 0.25 to 0.5 m by 2100. Globally, soil salinity will lead to higher food costs, more food shortages, and poorer yields from many farmers, which means reduced income.

4.2.3 Impacts of Climate Change on Evapotranspiration

Climate change has an impact on evapotranspiration (ET), one of the key components of the hydrologic cycle. The weather is changing over long periods of time, which is a sign of climate change (IPCC 2014). Evapotranspiration rates will fluctuate as a result of a shift in climatic factor patterns brought on by climate change (Helfer et al. 2012). Potential evaporation (ET_P), which measures the combined impact of numerous meteorological parameters, including solar radiation, wind speed, ambient temperature, vapor pressure, and humidity, can be thought of as a measure of atmospheric evaporative demand (Dinpashoh et al. 2018). On both a regional and a global level, atmospheric temperature is regarded as the most widely utilized indicator of climatic change. According to the IPCC report (2013), between 1913 and 2012, the average global temperature increased by 0.91 °C (Stocker et al. 2014). This increase in temperature is expected to continue throughout the twenty-first century, changing the hydrological cycle by changing both precipitation and evaporation (Huntington 2006). Water availability, quality, and quantity may be significantly impacted by this, especially in underdeveloped semiarid locations (Dastorani and Poormohammadi 2012). Potential evapotranspiration (ET_P), which describes a significant water loss from catchments, is widely acknowledged as a crucial hydrological variable. It can be used to calculate actual evapotranspiration (ET_a), scheduled irrigation, and other management methods in the crop field. It is related to groundwater recharge, runoff, and water movements in soil, some key forms of hydrological processes (Zhang et al. 2011; Xu and Li 2003). Climate change will likely result in more drought conditions by increasing potential evapotranspiration and increasing crop water use in areas with limited water resources (Thomas 2008). Climate change models predict that ET_P will rise during the next few years as a result of anticipated temperature acceleration (Goyal 2004; Liu et al. 2018).

4.2.4 Impact of Climate Change on Plants and Animals

Due to escalated temperatures due to climate change, shifts in life cycles are expected (e.g., early flower blooming, birds hatching earlier in spring, and many species shifting their migration pattern). Due to global warming, plants and animals will migrate toward poles, and the species which may fail to migration shall perish. Some scientists opine that 20–50% of the species could be committed to extinction with 2–3 °C of further warming.

4.2.5 Impact of Climate Change on Weather

Polar regions of Northern Hemisphere will heat up more than other areas of the planet, glaciers will melt, and sea ice will shrink. Winter and night temperatures will tend to rise more than summer and daytime temperatures. Warmer world will be more humid (because of more water evaporating from oceans). Water vapor is a GHG and induces warming phenomenon, but, on the other hand, more water vapor will produce more clouds which reflect sunlight back. More clouds may cause flash floods due to probability of cloud bursts.

Contrarily, enhanced greenhouse effect will lead to more evaporation of soil water, therefore, droughts are projected to become more intense. Weather patterns are expected to be less predictable and more extreme.

4.2.6 Impact of Climate Change on Health

Diseases like malaria, dengue fever, yellow fever, Japanese encephalitis, allergies, and respiratory diseases shall proliferate due to high temperature.

4.3 Climate Change Impacts on Aquatic Ecosystem

4.3.1 Climate Change Effect on the Physical, Chemical, and Biological Properties of Ocean

Ocean ecosystems are being significantly impacted by climate change, although the extent of these changes is still not fully understood. Despite the oceans' enormous capacity to absorb heat and carbon dioxide, the warming trend seems to be accelerating. More than 90% of the Earth's warming since 1950 has been observed in the oceans. As a result of climate change, ocean stratification has increased, ocean current regimes have changed, depleted oxygen zones have expanded, the geographic ranges of marine species have changed, and the growing seasons, diversity, and abundance of species communities have changed. As a result of atmospheric warming, inland glaciers and ice are melting, leading to rising sea levels that have a significant impact on shorelines (coastal erosion, salt water intrusion, and habitat destruction), as well as coastal human settlements. According to the IPCC, the global mean sea level will rise by 0.40 (0.26–0.55) m from 1986–2005 to 2081–2100 under low emission scenarios and by 0.63 (0.45–0.82) m under high emission scenarios. Furthermore, growing GHG emissions are expected to result in an increase in the frequency of extreme El Niño episodes.

4.3.2 Changes in Physical Properties

Oceanic circulation, sea-level rise, variations in water temperature, and intensified storms are few examples of how ocean physical features are changing.

4.3.3 Changes in Water Temperature

Although the ocean has absorbed more than 80% of the heat that climate change has added to the Earth system, the ocean is suffering as a result. Oceanic heat waves have occurred more frequently, by more than 50%, in the last century or so. A marine heat wave is described as having temperatures that are significantly higher than average for at least 5 days, brought on by changing warm currents and heat from blazing sunshine. Globally, there were, on average, more maritime heat wave days per year between 1982 and 2016 than to the prior period. In general, marine habitats are harmed by heat stress and heat waves. Heat waves have significant negative consequences on the environment and the economy, including coral bleaching, mass extinctions of marine species owing to heat stress, kelp forest loss, species migration, and the resulting changes in community structure. According to certain research, fish and mobile invertebrates may seek out uninhabited environments to escape heat waves, which in turn promote diversity. Due to fluctuations in the availability of prey and corals' vulnerability to bleaching at high temperatures, birds and corals both experience poor prognoses. Corals and sea grasses, which frequently serve as habitats and sources of food for numerous other animals, are hardly affected, which causes the negative impacts to spread throughout the ecosystem.

4.3.4 Melting of the Polar Ice

Polar ice is melting as a result of rising atmospheric temperature, and over the past 30 years, both the thickness and coverage of sea ice in the Arctic have significantly changed. Studies suggest that, between 1980 and 2008, the thickness of sea ice decreased by 50% to 1.75 meters and, between 1980 and 2008 (28 years), the expanse of sea ice decreased by an average of 11%, with evidence of a recent acceleration.

4.3.5 Rising Sea Levels

As compared to the past 2000 years, sea levels have risen, according to data from monitoring programs for the sea level and other sources. When the water temperature in the sea rises, the sea expands. Sea levels rise as a result of glacier and polar ice melting. Sea-level rise is also a result of human actions such the draining of wetlands, groundwater extraction, building of dams, and changes in land use. Since 41% of the world's population lives within 100 kilometers of the coast, sea-level rise is a huge concern. Thus, erosion of beaches and dunes is more prone to occur. Small islands where land is only few meters above sea level will experience salt water intrusion, for example, Tuvalu Island in western Pacific Ocean and Kiribati Island in Central Pacific Ocean. Likewise, the Netherlands (Northwestern Europe), Cyprus, as well as other countries around the "Mediterranean Basin" may need to spend a huge amount of money to protect its shorelines, whereas poor countries like Bangladesh may be forced to simply abandon low-lying coastal areas. The IPCC estimated figures point out that sea-level rise in 2100 was lower than the present rate of rising (3.1 mm per year). These rates, however, vary geographically and are not comparable worldwide.

4.3.6 Changes to the Ocean's Major Current Systems

Oceanic currents will be impacted and altered by changes in ocean temperature and wind patterns. Changes in the primary current systems in the ocean will have a huge impact on the global climate because the ocean currents are crucial in regulating the Earth's temperature. Due to a rise in sea surface temperature and ice melting, oceanographers have noticed alterations in the North Atlantic Ocean currents. The management of the world's ocean currents relies heavily on the Atlantic. The Southern and Pacific oceans experience currents due to the vast amounts of cooler water that sink in this region. Therefore, a slowdown of the currents in this area has effects on the entire world. The North Atlantic storms intensify, the entire Northern Hemisphere cools, the Indian and Asian monsoon regions dry up, and reduced ocean mixing causes a decrease in plankton and other marine life. Additionally, it would cause the Southern Hemisphere to warm up. The IPCC came to the conclusion that if the temperature rises by 4 °C and GHG emissions continue to rise, circulation may decrease by up to 54% by this century.

4.3.7 Changes in Chemical Properties

By absorbing significant amounts of CO_2, the ocean serves as a carbon sink. Due to the fact that CO_2 is instantly reactive in sea water, the ocean's capacity to absorb CO_2 is ten times greater than that of freshwater. The ocean's chemical composition alters as a result of this process.

4.3.8 Ocean Acidification

The oceans are said to absorb around one-third of the CO_2 produced by human activity and released into the atmosphere. The concentrations of the hydrogen carbonate (HCO^{3-}) and carbonate (CO_3^{2-}) ions change as soon as CO_2 from the atmosphere reaches the water because it reacts with water molecules to generate carbonic acid. The ocean has become more acidic as a result, endangering the existence of many marine species and ecosystems, yet this has greatly slowed down global warming. It is discovered that the current ocean acidification is 30 times more severe than the natural variation. The average ocean surface pH has also declined since the Industrial Revolution by around 0.1 unit, which is noteworthy because it results in a 25% rise in acidity. By year 2100, it is predicted that ocean acidification levels will have increased by 144% if CO_2 emissions continue. The ability of marine organisms like corals to build calcium carbonate shells is drastically reduced by higher acidity. Studies have demonstrated that the creation of calcium carbonate is being hampered by ocean acidification. Ocean acidification also worsens current "physiological stresses" and significantly lowers the growth and survival rates of a small number of marine species, especially in the early phases of development.

4.3.9   Hypoxia

Ocean water warms as its heat content rises, which reduces the water's ability to hold dissolved oxygen. The prevalence of low oxygen levels (hypoxia), which makes marine ecosystems more vulnerable, is observed to increase as a result of regional and global climate change as well as coastal eutrophication. The term "hypoxic waters" refers to water with an oxygen content of less than 2 ppm. Additionally, as surface water is warmer, it no longer mixes as much with the ocean's depths. Reduced mixing of warmer, lighter surface water with denser, deeper water obstructs the delivery of dissolved oxygen to aquatic creatures that live deep in the water. This may result in "oxygen minimum zones" where it would be difficult for plants, fish, and other species to thrive. The Gulf of Mexico, the Baltic, the Adriatic, the East China Sea, and the northwestern shelf of the Black Sea are a few well-known examples of these "dead zones." It is a growing issue that has serious effects on marine life, such as changing their habitats and behaviors, causing death, and causing catastrophic alterations.

4.3.10   The Vulnerability of Marine Organisms

4.3.10.1   Coral Bleaching

About one-third of all marine animals have been found to live in coral reefs, making them a significant group of marine life. A number of variables, including temperature (the ideal range is between 22° and 29 °C), nutrients, currents, turbidity, light, pH, calcium carbonate concentration, etc., affect the growth and development of coral reefs. In terms of climate change, coral reefs are vulnerable to temperature increases. Corals expel the algae (zooxanthellae) residing symbiotically in their tissues when the water is warm (warmer than the ideal temperature for coral growth), which gives corals their bleached appearance popularly known as coral bleaching. The corals eventually die as a result of coral bleaching.

Acidification also has a negative impact on coral skeletons by slowing down coral calcification, which results in poor skeleton density and corals that are more prone to breaking. However, the effects of acidification vary depending on the species, possibly because different organisms have varying degrees of control over the pH of their calcification sites. The survival of numerous marine species is also under danger due to rising sea levels. Species that depend on relatively shallow water for photosynthesis, including corals and sea grass meadows, are also in risk. Rising sea levels have an impact on a number of marine species, for example, the Hawaiian monk seal. There is a 4% annual decline in the number of monk seals, according to reports. Numerous marine creatures may be impacted by declining krill and phytoplankton populations.

4.3.11 Vulnerability and Impact Assessment of Wetlands to Climate Change

The hydrological regimes, specifically the nature and unpredictability of the hydroperiod and the frequency and severity of extreme events, may be where climate change has the most noticeable impact on wetlands. However, other climate-related factors, such as elevated temperatures and altered evapotranspiration, altered biogeochemistry, altered levels and patterns of suspended sediment loadings, fire, oxidation of organic sediments, and the physical effects of wave energy, may be significant in determining regional and local impacts. Changes in hydrology, the direct and indirect effects of climate changes, and land use change are expected to act as mediators of pressures on wetlands. According to STRP (2002), changes in base flows could have the following effects: altered hydrology (depth and hydroperiod); increased heat stress in wildlife; expanded range and activity of some increased flooding, landslide, pest, and disease vectors; increased soil erosion; increased tropical cyclone activity, avalanche, and mud slide damage; increased flood runoff resulting in a decrease in recharge of some floodplain aquifers; and decreased water availability. Wetland systems are prone to changes in the quantity and quality of the water supply, and they are particularly susceptible to such changes. Individual wetland habitats' hydrology will be impacted by climate change primarily through modifications to the highly variable global precipitation and temperature regimes. The effects of climate change will vary according to different temperature and precipitation regimes, and so will the restoration strategies, given the diversity of wetland types and their unique characteristics. Ocean animals and habitats are impacted by both ocean acidification and rising ocean temperatures. Oceanic mammals, fish, and seabirds are all in grave danger due to climate change, which will result in mass extinctions, greater death rates, and the loss of reproductive habitats for many species. The metabolism of individuals, species life cycles, connections between predators and prey, and habitat modification are all impacted by physiochemical changes in the properties of sea water.

References

Al-Najar H, Ashour EK (2012) The impact of climate change and soil salinity in irrigation water demand in the Gaza strip. J Water Clim Change 4:118–130

Dastorani MT, Poormohammadi S (2012) Evaluation of the effects of climate change on temperature, precipitation and evapotranspiration in Iran. In: International Conference on Applied Life Sciences (ICALS2012), Turkey, 2012, September 10–12

Dinpashoh Y, Jahanbakhsh-Asl S, Rasouli AA, Foroughi M, Singh VP (2018) Impact of climate change on potential evapotranspiration (case study: west and NW of Iran). Theor Appl Climatol 136:185–201

Goyal R (2004) Sensitivity of evapotranspiration to global warming: a case study of arid zone of Rajasthan (India). Agric Water Manag 69:1–11

Gregorich EG, Carter MR, Angers DA, Monreal CM, Ellert BH (1994) Towards a minimum data set to assess soil organic matter quality in agricultural soils. Can J Soil Sci 74:367–385

Helfer F, Lemckert C, Zhang H (2012) Impacts of climate temperature and evaporation from a large reservoir in Australia. Hydrology 475:365–378

Huntington TG (2006) Evidence for intensification of the global water cycle: review and synthesis. J Hydrol 319:83–95

IPCC (2013) Summary for policymakers. In: Stocker TF, Qin D, Plattner GK, Tignor M, Allen SK, Boschung J, Nauels A, Xia Y, Bex V, Midgley PM (eds) Climate change 2013: the physical science basis. Contribution of working group I to the fifth assessment report of the intergovernmental panel on climate change. Cambridge University Press, Cambridge/New York

IPCC (2014) Summary for policymakers. In: Edenhofer O, Pichs-Madruga R, Sokona Y, Farahani E, Kadner S, Seyboth K, Adler A, Baum I, Brunner S, Eickemeier P, Kriemann B, Savolainen J, Schlomer S, von Stechow C, Zwickel T, Minx JC (eds) Climate change, 2014, mitigation of climate change. 30 p. Contribution of working group III to the fifth assessment report of the intergovernmental panel on climate change. Cambridge University Press, Cambridge

Lal R (2011) Soil health and climate change: an overview. In: Soil health and climate change. Springer, Berlin/Heidelberg, pp 3–24

Liu Q, Yan C, Ju H, Garré S (2018) Impact of climate change on potential evapotranspiration under a historical and future climate scenario in the Huang Huai-Hai Plain, China. Theor Appl Climatol 132:387–401

Ozturk M, Mermut A, Celik A (2011) Land degradation, urbanisation. Land Use & Environment NAM S & T (Delhi-India), 445 pp

Stocker T, Qin D, Plattner GK, Tignor M, Allen SK, Boschung J, Nauels A, Xia Y, Bex V, Midgley PM (2014) Climate change 2013: the physical science basis. Cambridge University Press, Cambridge/New York

STRP (Scientific and Technical Review Panel of the Ramsar Convention on Wetlands) (2002) New guidelines for management planning for Ramsar sites and other wetlands. "Wetlands: water Life, and culture" 8th meeting of the conference of the contracting parties to the convention on wetlands (Ramsar, Iran, 1971) Valencia, Spain, 18–26 November 2002

Szabolcs I (1990) Impact of climatic change on soil attributes: influence on salinization and alkalinization. Dev Soil Sci, Elsevier 20:61–69

Teh SY, Koh HL (2016) Climate change and soil salinization: impact on agriculture, water, and food security. Int J Agric Forest Plant 2:1–9

Thomas A (2008) Agricultural irrigation demand under present and future climate scenarios in China. Glob Planet Chang 60:306–326

Weil RR, Magdoff F (2004) Significance of soil organic matter to soil quality and health. In: Weil RR, Magdoff F (eds) Soil organic matter in sustainable agriculture. CRC Press, Boca Raton, pp 1–43

Xu Z, Li J (2003) A distributed approach for estimating catchment evapotranspiration: comparison of the combination equation and the complementary relationship approaches. Hydrol Process 17:1509–1523

Zhang Q, Xu C, Chen X (2011) Reference evapotranspiration changes in China natural processes or human influences. Theor Appl Climatol 103(479):488

Chapter 5
Carbon Capture and Storage

5.1 Introduction

One of the most important elements of the physical environment and the living world is carbon. As one of the main constituents of all cellular life forms and the precursor of all metabolic processes, carbon serves as the foundation of all living systems in the world of living things. Additionally, it contributes significantly to keeping the Earth's average temperature at the ideal level of 15 °C through the greenhouse effect, which is necessary for life to flourish. The CO_2, CH_4, and N_2O gases are significant contributors to the anthropogenic greenhouse effect. CO_2 plays a leading role among all of these greenhouse gases because it accounts for 60% of the entire greenhouse impact (Bhardwaj and Panwar 2003).

Capturing and storing carbon for a long time can be a solution to global warming. The most prevalent instance of this occurs in nature when trees and plants use photosynthesis to store carbon as they take in carbon dioxide (CO_2) to grow (Fig. 5.1).

Trees and plants are crucial components in efforts to slow down global warming because they absorb carbon that would otherwise retain heat (infrared wavelengths) in the atmosphere. According to Sedjo and Sohngen (2012), carbon sequestration through biomass and soil is a long-term technique of capturing and storing atmospheric CO_2 and will prove to be a practical and affordable solution to reduce the rising levels of greenhouse gases. Various sequestration strategies and methodologies are presently available which are discussed in this chapter.

© The Author(s), under exclusive license to Springer Nature Switzerland AG 2023
M. A. Dervash et al., *Phytosequestration*, SpringerBriefs in Environmental
Science, https://doi.org/10.1007/978-3-031-26921-9_5

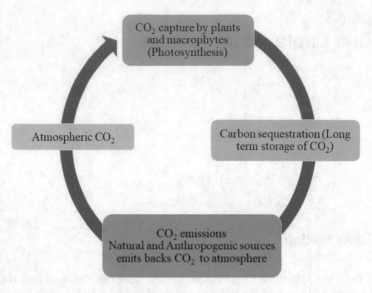

Fig. 5.1 CO_2 cycle and carbon sequestration

5.2   Strategies to Sequester Atmosphere Carbon Dioxide

Numerous natural and man-made processes have the ability to behave as atmospheric sponges, absorbing CO_2. As a result, descriptions of various carbon sequestration techniques are as follows:

1. *Glacial storage*: Solid CO_2 clathrates can be used. The clathrates are stored in columns of water inside glaciers which are then refrozen. Issues in this method of storage are with remoteness of areas.
2. *Biogenic methane*: In this method, CO_2 is injected into depleted coal seam; then, biological/geochemical processes are used to convert CO_2 to CH_4. This is an example of closed-loop fuel systems, thus CO_2 recycling.
3. *Mineralization*: CO_2 is combined with minerals from various sources to create stable carbonate compounds:

$$\text{Flue gas}\left(CO_2\right) + Ca\,/\,Mg\,/\,Na\,/\,Fe\,/\,Hg \rightarrow \text{Carbonates}\left(\text{SequesTech}\right).$$

Where: "Ca = calcium, Mg = magnesium, Na = sodium, Fe = iron, and Hg = mercury"

4. *Enzymatic catalysis*: In this technique, activity of carbonic anhydrase enzyme is exploited:

$$CO_2 \rightarrow \text{Energy Crops} \rightarrow \text{Fermentation} \rightarrow \text{Distillation} \rightarrow \text{Ethanol}$$
$$\downarrow$$
$$CO_2 \text{ captured and stored}$$

Natural Sinks

There are five distinct global carbon pools, according to a thorough study on carbon sinks by Schrag (2007):

(I) *Oceanic pool*: It is considered to be the largest pool at 38,000 petagrams $(Pg = 10^{20} \text{ g})$, and it is growing at a pace of 2.3 Pg C yr^{-1}.

(II) *Geological carbon pool*: Also known as the fossil fuel pool, it is thought to be 4130 Pg in size, with coal accounting for 85%, oil for 5.5%, and gas for 3.3%. There are 678 Pg of coal (3.2 Pg yr^{-1} of production), 146 Pg of oil (3.6 Pg yr^{-1} of production), and 98 Pg of natural gas in verified fossil fuel reserves (1.5 Pg yr^{-1} of production). Therefore, it is clear from this account that the geological pool is being depleted at a rate of 7.0 Pg C yr^{-1} due to the combustion of fossil fuels.

(III) *Soil pool*: Being the third biggest active carbon pool, soils have a sizable and dynamic carbon pool (Eshel and Singer 2007). There is a chance to increase the amount of carbon stored in these soils by well-planned land use management, as it has been noted that soil carbon pools have drastically decreased over the past several centuries. Carbon in soils must typically be managed indirectly through manipulation of plants and nutrients, but carbon in vegetation can be managed unwaveringly through land use.

(IV) *Atmospheric pool*: The atmospheric pool, which contains 760 Pg of CO_2, is the fourth significant pool. It is increasing at a rate of 3.5 Pg C yr^{-1} or 0.46% per year.

(V) *Terrestrial carbon pool*: Terrestrial carbon pool, which is believed to be around 2860 Pg, is together made up of the soil and biotic carbon pools. The biotic pool is the smallest of the global carbon pools, with an estimated mass of 560 Pg. The oceanic pool, which absorbs 92.3 Pg yr^{-1} and releases 90 Pg yr^{-1} with a net positive balance of 2.3 Pg C yr^{-1}, is likewise connected to the atmospheric pool. Additionally, it is anticipated that by 2100, the oceanic pool will absorb almost 5 Pg C each year (Orr et al. 2001). Thus, the terrestrial ecosystem is one of the primary natural CO_2 scrubbers that can be considerably increased through careful management. The only realistic method of removing significant amounts of CO_2 from the atmosphere involves absorbing it, integrating it into the biological system and biomass of the flora, and then transferring it into the soil. As a result, carbon is stored in the plants, where some of it is later partially degraded by microbial activity to form organic carbon in the soil. The arboreal regions eventually serve as a "sink" for this trapped carbon.

5.3   Carbon Sequestration by Vegetation

Through the process of photosynthesis, algae, forests, oceans, and most other plants accumulate carbon. A healthy ecosystem's ability to absorb atmospheric carbon is known as a "carbon sink." As part of photosynthesis, every living plant matter absorbs CO_2, but due to their size and extensive root systems, trees consume far more CO_2 than smaller plants. In essence, trees are the most effective "carbon sinks" in nature since they are the kings of the plant kingdom and have a lot more "woody biomass" to store CO_2 than lower plants.

Carbon sequestration through vegetation can be quantified by two broader sampling methodologies:

5.3.1   Destructive Sampling Method

This method necessitates complete plant uprooting in order to calculate the total tree biomass and associated carbon storage. The total biomass of a tree is calculated as the sum of the biomass of its stem, branches, leaves, and roots. Using the tissue-specific carbon content as measured by the Ash Content Method recommended by Negi et al. (2003), it is possible to calculate the standing carbon density (above- and below-ground) of plants, such as *Salix alba*:

$$\text{Carbon}(\%) = 100 - \left(\text{Ash weight} + \text{moecular weight of O}_2\left(53.3\right)\text{in C}_6\text{H}_{12}\text{O}_6\right)$$

The carbon density (tons ha^{-1} or Kg tree^{-1}) is then determined using Hu's approach (2006):

$$C_b = \alpha \times M$$

Where:
C_b is the carbon density (above- and below-ground).
α is the average carbon content (above- and below-ground biomass).
M is the total biomass (tons ha^{-1} or Kg tree^{-1}).

Then, using the method described by Dury et al. (2002), the amount of elemental carbon (CO_2) taken from the atmosphere is computed:

$$CO_{2e} = C_b \times 3.67$$

5.3.1.1   Soil Organic Carbon

On the other hand, Walkley and Black's approach (1934) is used to estimate soil organic carbon.

The Van Bemmelen factor of 1.724 is used for conversion of organic carbon content to organic matter content (%).

$$\text{Organic matter content}\,(\%) = \text{Organic carbon} \times 1.724$$

$$C_s = \frac{100\sum(D \times d \times O)}{1.724}$$

The soil organic carbon density is then determined as per procedure followed by Hu (2006):

$$C_s = \frac{100\sum(D \times d \times O)}{1.724}$$

Where:
C_s is the soil carbon density (tons ha^{-1} or Kg tree^{-1}).
D is the soil depth (cm).
d is the soil bulk density (g cm^{-3}).
O is the soil organic matter content (%).

5.3.1.2   Herb and Litter Biomass Estimation

Herb measurements can be obtained by gathering all the herbs and recently fallen litter from (1 m × 1 m quadrate) herb plot. Fresh weight samples are taken right away after detaching the soil that is adhered to the roots. Then, 200-g-sized sub-samples from the overall sample are removed, placed in paper bags, marked, and brought to the laboratory for analysis. The formula provided by Anup (2013) is used to compute the biomass of plants and litter per unit area:

$$LHG\left(\text{t ha}^{-1}\right) = \frac{W_F \times W_{SS,\text{dry}}}{Z \times W_{SS,\text{wet}}}$$

Where:
LHG is the biomass of grass, herbs, and leaf litter (t ha^{-1}).
W_F is the weight of the fresh field sample of grass, plants, and leaf litter that was taken destructively within Z (g).
Z is the size of the area used to gather grass, herbs, and leaf litter (m^2).
$W_{SS,\text{dry}}$ is the weight of the oven-dried sub-sample of grass, plants, and leaf litter brought to the laboratory to ascertain its moisture content (g).
$W_{SS,\text{wet}}$ is the weight of the fresh sub-sample of grass, herbs, and leaf litter that needs to be transported to the laboratory for assessment of moisture content (g).

According to MacDicken (1997), a root-to-shoot ratio of 1:5 is utilized when calculating below-ground biomass. The total weight (oven-dried t ha-1) is multiplied by the global factor 0.50 conversion factor to determine the carbon stock in under-storey vegetation (Malhi et al. 2004). The formula is then used to calculate the total carbon:

$$\text{Carbon}\left(t\,ha^{-1}\right) = \text{Biomass}\left(t\,ha^{-1}\right) \times \text{Carbon}\left(\%\right)$$

5.3.1.3   Soil Microbial Biomass Carbon

The Vance et al. (1987) recommended chloroform fumigation extraction method to determine the soil microbial biomass carbon by quantifying the difference in extractable organic carbon between fumigated and unfumigated soils:

$$MBC\left(\mu g\,g^{-1}\right) = 2.64 \times E_{oc}$$

Where E_{oc} is the acronym for extractable organic carbon and 2.64 is the proportionality factor for the biomass C released during fumigation extraction.

5.3.1.4   Net Carbon Sequestration

The formula recommended by Dury et al. (2002) is used to compute net carbon sequestration (CS ton ha^{-1}):

$$CS = \text{Change in } TR + \text{Change in L} + \text{Change in S} + \text{Change in M}$$

Where:
   TR is the annual change in shoot and root biomass (above- and below-ground).
   L is the annual change in litter biomass.
   S is the annual change in soil organic carbon.
   M is the annual change in microbial biomass.

5.3.2   *Non-destructive Sampling Method*

In this method, regression general volume equations (allometric equations for various tree species) involving "(i) volumetric equations developed by Forest Survey of India and Forest Research Institute (FSI 1996), and (ii) literature on specific gravity developed by Indian Council of Forestry Research and Education Dehradun (Rajput et al. 1996)" are used.

The diameter at breast height (DBH at approximately 1.3 m) is used to calculate the biomass. For measuring, tree species with DBH > 10 cm are considered. Using a measuring tape and a Ravi multimeter, respectively, the girth at breast height (GBH) and tree height (h) are determined. Using a portable GPS unit, the location of the quadrates is recorded, as well as other site variables like slope, aspect, altitude, and tree density.

Volume is calculated using DBH and height using suggested volume equations (FSI 1996), which, as per Rajput et al. (1996), is further converted into above-ground biomass (AGB) using specific gravities of the respective species. Commercial bole biomass is converted into total above-ground biomass using biomass expansion factors (Brown 1997). Using a factor of 0.26, below-ground biomass is computed from above-ground biomass (Cairns et al. 1997). A conversion factor of 0.50 is used to convert the estimated total biomass into carbon equivalent (Guo et al. 2010) (Table 5.1; Fig. 5.2).

5.3.2.1   Soil Organic Carbon Density and SOC CO_2 Mitigation

According to IPCC Good Practice Guidance for Land Use, Land-Use Change, and Forestry (LULUCF), the following equation should be used to determine SOC density (t ha^{-1}) for (O) horizon (0–30 cm):

Table 5.1 General volume equations for few tree species as per FSI (1996)

Test species	Volume equations	Specific gravity (g cm^{-3})
Salix alba	"V = 0.193297 − 2.267002 D + 10.679492D^2"	0.37
Populus deltoides	"V = 0.193297 − 2.267002 D + 10.679492D^2"	0.40
Platanus orientalis	"V = 0.193297 − 2.267002 D + 10.679492D^2"	0.48
Aesculus indica	"$\sqrt{V}$ = 0.220191 + 3.923711 D − 1.117475 $\sqrt{D}$"	0.39
Robinia pseudoacacia	"V = 0.193297−2.267002 D + 10.679492D^2"	0.629
Quercus spp.	"$\sqrt{V}$ = 0.240157 + 3.820069 D − 1.394520 $\sqrt{D}$"	0.692
Morus alba	"$\sqrt{V}$ = − 0.07109 + 2.99732 D − 0.26953 $\sqrt{D}$"	0.577
Parrotia spp.	"$\sqrt{V}$ = − 0.07109 + 2.99732 D − 0.26953 $\sqrt{D}$"	0.758
Celtis australis	"V = 0.193297−2.267002 D + 10.679492D^2"	0.44
Prunus armeniaca	"V = 0.193297−2.267002 D + 10.679492D^2"	0.513
Pinus wallichiana	"V = 0.193297−2.267002 D + 10.679492D^2"	0.427
Juglans regia	"$\sqrt{V}$ = 0.207299 + 3.254007 D"	0.629
Mixed broadleaved	"V = 0.193297−2.267002 D + 10.679492D^2"	0.75 (*Malus domestica*) 0.53 (*Rhus* spp.)

Fig. 5.2 Carbon sequestration through vegetation (*Platanus orientalis*)

$$SOC = \sum_{\text{Horizon1}}^{\text{Horizon}n} SOC_{\text{Horizon}}$$

$$= \sum_{\text{Horizon1}}^{\text{Horizon}n} \{(SOC) \times \text{Bulk density} \times \text{depth} \times (1-C) \times 100\}_{\text{horizon}}$$

Where:

SOC is the representative soil organic carbon content for the forest type, tons (t) C ha^{-1}.

SOC horizon is the soil organic carbon content for a constituent soil horizon, t ha^{-1}.

(SOC) is the concentration of SOC in a given soil mass obtained from analysis, g C (kg soil)$^{-1}$.

Bulk density is the soil mass per sample volume, t m^{-3}.

Depth is the thickness of soil layer, m.

C fragments is the % volume of coarse fragments/100.

After being multiplied by a factor of 3.67 (C equivalent of CO_2), the SOC density calculated for each stratum is then converted to SOC CO_2 mitigation. The results would reveal how much CO_2 is absorbed by the soil beneath each stratum.

References

Anup KC, Govinda B, Ganesh RJ, Suman A (2013) Climate change mitigation potential from carbon sequestration of community forest in Mid Hill region of Nepal. Int J Environ Protect 3.33–40

Bhardwaj SD, Panwar P (2003) Global warming and climate change-effect and strategies for its mitigation. Indian Forest 32:741–748

Brown S (1997) Estimating biomass and biomass change of tropical forests: a primer. In: FAO Forestry Papers Vol. 134. Food and Agriculture Organization of the United Nations, Rome, Italy. http://www.fao.org/icatalog/search/dett.asp?aries_id=7736

Cairns MA, Brown S, Helmer EH, Baumgardener GA (1997) Root biomass allocation in the world's upland forests. Oecologia 111:1–11

Dury SJ, Polglase PJ, Vercose T (2002) Greenhouse resource kit for private forest growers. Commonwealth Department of Agriculture and Forestry, Canberra. IV: 95

Eshel G, Singer MJ (2007) Total soil carbon and water quality: an implication for carbon sequestration. Soil Sci Soc Am J 71:397–405

FSI (1996) Volume equations for forests of India, Nepal and Bhutan. Forest Survey of India, Ministry of Environment and Forests, Govt. of India, Dehradun

Guo Z, Fang J, Pan Y, Birdsey R (2010) Inventory based estimates of forest biomass carbon stocks in China: a comparison of three methods. For Ecol Manag 259:1225–1231

Hu J (2006) Carbon storage of artificial forests in rehabilitated lands in the upper reaches of the yellow river. Front Forest China 1:268–276

MacDicken K (1997) A guide to monitoring carbon storage in forestry and agroforestry projects, Forest Carbon Monitoring Programme, Winrock. International Institute for Agricultural Development, Arlington, p 87

Malhi Y, Baker TR, Phillips OL, Almeida S, Alvarez E, Arroyo L, Chave J, Czimczik CI, Di FA, Higuchi N, Killeen TJ, Laurance SG, Laurance WF, Lewis SL, Montoya LM, Monteagudo A, Neill DA, Nunez VP, Patino S, Pitman NCA, Quesada CA, Silva JNM, Lezama AT, Vasquc MR, Terborgh J, Vinceti B, Lloyd J (2004) The above-ground coarse wood productivity of 104 Neotropical forest plots. Glob Change Biol 10:563–591

Negi JDS, Manhas RK, Chauhan PS (2003) Carbon allocation in different components of some tree species of India: a new approach for carbon estimation. Curr Sci 85:1528–1531

Orr JC, Maier-Reimer E, Mikolajewicz U, Monfray P, Sarmiento JL, Toggweiler JR, Taylor NK, Palmer J, Gruber N, Sabine CL, Quere CL, Key RM, Boutin J (2001) Estimates of anthropogenic carbon uptake from four three-dimensional global ocean models. Global Biogeochem Cy15:43–60

Rajput SS, Shukla NK, Gupta VM, Jain JD (1996) Timber mechanics: strength classification and grading of timber. Indian Council of Forestry Research and Education Publication, Dehradun, p 103

Schrag DP (2007) Preparing to capture carbon. Science 315:812–813

Sedjo R, Sohngen B (2012) Carbon sequestration in forests and soils. Annu Rev Resour Econ 4:127–144

Vance ED, Brookes PC, Jenkinson DS (1987) An extraction method for measuring soil microbial biomass. Soil Biol Biochem 19:703–707

Walkley AE, Black JA (1934) An examination of the Degtjareff method for determining soil organic matter and proposed modification of the chromic acid titration method. Soil Sci 37:29–38

Chapter 6
Future Climate Through the Window of Climate Models

6.1 Introduction

In recent years, there has been a growing concern about climate change all across the globe. The increasing average global temperatures, the rising of sea level and submerging of low-lying islands, the loss of biodiversity, and shrinking of hotspots all over the world have led to increasing distress about the changing climate (Ozturk et al. 2015). Historically speaking, ever since the origin of the Earth, it is not the first time that the climate is changing; however, the enhanced anthropogenic contribution to climate change is what makes the entire scenario worrisome. As a result, scientists all over the world are thinking of various strategies to understand the phenomenon of climate change, to gain an insight about its implications, and also to quantify the changing climate by means of models. It is where the role of climate models comes into picture. These models attempt to quantify the changing climate by using physical laws of radiation and energy and studying the radiation behavior and flux at the surface of the Earth. So, there may be regional or global climate models to quantify the extent of change that the Earth is undergoing in terms of climate and also include clouds and aerosols in more complex models for understanding their dual role in radiative forcing of the Earth that eventually leads to climate change. Various emission scenarios of greenhouse gases are also discussed in this chapter that has been reported by the Intergovernmental Panel on Climate Change (IPCC).

M. A. Dervash et al., *Phytosequestration*, SpringerBriefs in Environmental Science, https://doi.org/10.1007/978-3-031-26921-9_6

6.2   Analogues from Past Climate

As per the IPCC report (2007), rapid climate changes have been observed in the recent past and are further expected to increase in the near future. The analogue approach is a novel method of testing the ground realities to the outputs of climate models. In this approach, the analogue tool connects the sites that are analogous (similar) to climates at sites across other geographic locations, implying climates across space as well as time (means with respect to historical or projected future climates). For example, if any place in the world has the present climate that is similar to the future predicted climate elsewhere, then these two sites can provide interesting observations on adaptation strategies to be followed for mitigating or at least minimizing the adverse effects of predicted climate change in the future. This approach is usually helpful for complex systems that face difficulty in climate models. But this approach holds true only if climate is the driving force behind the observations in differences at the two sites. In a holistic way, the analogue approach can help relate global models with targeted field studies (Ramirez-Villegas et al. 2011) to explore mechanism of adaptation. Such comparisons between sites can be used to enable farmers to develop a knowledge framework for realizing the future of site-specific agricultural output. When analyzing case studies where the success or failure of an adaptation mechanism may be determined, the analogue methodology can be used to comprehend historical data. For future-related analyses, users in this method supply a location known as the reference location, variables like rainfall and temperature, one or more climatic scenarios (e.g., 2020, 2030, or 2050), an SRES emission scenario (IPCC 2000; Moss et al. 2010), and a global climate model (GCM) (IPCC 2007; PCMDI 2007). Including daily or monthly data, the temporal data for variables extends from an hourly to annual basis. The analogue tool would first utilize and forecast the climate scenario and then compare the current climate for all locations where data available with A's projected climate in order to locate suitable global present-day analogues for the 2050 climate of a specific reference site A. The dissimilarity index is then used to compare the outcome. In addition to climate, additional factors like soils, crops, and socioeconomic traits are taken into account. Depending on the volume of data, computing resources, and spatial resolution of the data, the outputs are generated for any geographic region at any resolution equal to or above 1 km.

6.3   Climate Models

A qualitative and/or quantitative representation of an object or phenomena is called a model. The term "general circulation model" (GCM) refers to computer programs that mimic the climate of the planet. Climate models simulate the past climate and forecast the future climate using physical characteristics and processes as input. Climate models are sometimes referred to as an extension of weather forecasting, with the main difference being that decades rather than hours are the dominant focus of time. The five elements of the climate system, atmosphere, hydrosphere, lithosphere, cryosphere, and biosphere, are mathematically represented by these models based on the physical and chemical laws of thermodynamics, fluid dynamics, radiative transfer, and biological interactions. Inclusion of clouds and the carbon cycle in climate models has also received attention recently. These models can therefore be used to mimic long-term variations in temperature, precipitation, winds, and oceanic circulations. However, inability arises due to these models to accurately simulate the Earth's complex and dynamic climate due to their reliance on assumptions and mathematical methods. However, these models have significantly improved recently to offer more insight into the upcoming environment. The ability of these models to recreate historical or present climates can be used to test their accuracy. In order to understand how the climate might change rather than how it will change based on objective and empirical data, climate models are a crucial tool. Normally, "scientists validate their models by comparing them against real-world observations or testing against past changes in the Earth's climate (hindcasts) such as temperature, rainfall, snow, hurricane formation, sea ice extent and many other climate variables. Also, it has been observed that the average of all models can be more accurate than most individual models in terms of higher reliability and consistency when several independent models are combined" (https://www.carbon-brief.org/qa-how-do-climate-models-work).

6.4 Types of Climate Models

The Intergovernmental Panel on Climate Change (IPCC) has created climate change scenarios using the best climate change models. The Canadian Climate Centre and the Hadley Centre in the United Kingdom have produced two main models which are used to project changes in the climate. Other weather simulations have been created at the National Center for Atmospheric Research, NOAA's Geophysical Fluid Dynamics Laboratory, NASA's Goddard Institute for Space Studies, and Max Planck Institute for Meteorology in Germany.

The output from these models varies depending on the inputs given to them as well as the uncertainties in greenhouse gas emissions, despite the fact that they operate on similar principles. For instance, the Canadian model predicts higher temperatures over the United States, but the Hadley model predicts a wetter climate. Since melting ice and snow in these locations reduce reflectivity and allow for greater heat absorption, practically all climate models generally expect a higher increase in temperature at places in medium to high latitudes. These models predict that the net effect of adding greenhouse gases and fossil fuel combustion results in producing a warmer climate IPCC anticipates a warming of 11 °C to 6.4 °C between 1900 and 2100. Thus, it becomes imperative to understand the differences in model projections for interpreting the results from the models. Some models have been discussed below:

6.4.1 Energy Balance Models (EBMs)

By taking into account the balance between incoming solar energy (insolation) and outgoing solar energy, which takes the form of heat emitted back into space, these models determine surface temperature as a variable for climate.

6.4.2 Zero-Dimensional Models

These models treat the entire planet as a single entity or, more precisely, as a single point. Consider a simple radiant heat transfer model that averages outgoing energy and treats the Earth as a single point.

6.4.3 One-Dimensional Models

These models include the transmission of energy across various latitudes of the Earth's surface in addition to treating the planet as a single entity. The key energy balance model compares the insolation to the energy that it emits back into space:

$$(1-\alpha)S\pi r^2 = 4\pi r^2 eST^4$$

Where:

$(1 - \alpha)S\pi r^2$ refers to the insolation.

$4\pi r^2 eST^4$ represents the outgoing energy from the sun calculated from Stefan-Boltzmann's constant using constant radiative temperature T.

S is the solar constant = 1367 W/m.

α is the Earth's average albedo (~0.3).

r is the Earth's radius, approximately 6.371×10^6 m.

e is the effective emissivity of earth, about 0.612.

S is Stefan-Boltzmann's constant, approximately 5.67×101 K + m^2 s.

T is the radiative temperature (K).

If we factor out πr^2, then

$$(1-\alpha)S = 4eST^4$$

The average Earth temperature is 288 K according to the equation above, which indicates the effective radiative temperature of the planet. These one-dimensional models do not examine the issue of temperature distribution on the Earth or the mechanisms responsible for energy circulation across the Earth. This model calculates the impact of changes in solar output, albedo, or the Earth's emissivity on surface temperature.

6.4.4   Radiative-Convective Models

These models are improvements to one-dimensional models that can predict how different greenhouse gas concentrations will affect emissivity and, consequently, surface temperature. Additionally, these models represent the convection of heat or radiative transmission of energy through the atmospheric altitudes. The temperature and humidity of various atmospheric layers can be calculated using radiative-convective models.

6.4.5   General Circulation Models (GCMs)

These four-dimensional (4-D) models, often known as global climate models, mimic the climate based on physical laws, the flows of air and water in the atmosphere and/or the seas, as well as the transfer of heat. They include time as a parameter. Earlier GCMs were only concerned with one aspect of the atmosphere, such as the seas or just the atmosphere.

The NOAA Geophysical Fluid Dynamics Laboratory created the first general circulation climate model that included both atmospheric and oceanic processes in the 1960s. Early GCMs only modeled a single part of the Earth system, such as the atmosphere or seas, but they did it in three dimensions, accounting for many kilometers of height in the atmosphere or oceanic depth in scores of model layers.

A comprehensive mathematical description of the main elements of the climate system is called a global climate model (GCM) whose salient features are mentioned below:

- The atmospheric element mimics aerosols and clouds.
- The features of the land's surface include flora, snowpack, and water bodies.
- The ocean component simulates biogeochemistry and current flow.
- The sea ice component controls water and air exchanges as well as the absorption of solar radiation.

The world is divided into a three-dimensional grid of cells in these climate models, and equations are calculated on the global grid for a set of climate variables like temperature as well as exchange fluxes of heat, water, and momentum for each component (atmosphere, land surface, ocean, and sea ice). The computing power and capability of the computer to solve these equations determine the grid size. When the resolution is good, more grid cells are needed, and when the grid cells are farther apart, less calculations are needed, though even the details aren't as detailed.

Furthermore, climate models include both simulated and parameterized processes. Simulated processes are larger than grid scale and are founded on principles like mass, energy, and momentum conservation. The model used to simulate tropical cyclone, and storm activity is an illustration of a simulated process. Parameterized processes employ both scientific concepts and empirical data to portray processes that are smaller than grid scale. A cloud and aerosol composition model is an illustration of a parameterized process.

The average climate for each grid cell is computed in order to parameterize climate models. The Earth is divided into grid cells. However, some processes, like the height of the landscape or the presence of clouds, take place at scales far lower than the grid size and might be missed. These variables are "parameterized" which denotes that they are defined in the computer code rather than being estimated by the model itself, to correct such mistakes in a model. The scattering of aerosols, snow cover, evaporation, condensation, soil characteristics, rain, surface roughness, and other factors are a few examples.

Since parameterized variables frequently cannot be reduced to a single value, estimation must be included in the model. To determine the value or range of values that enables the model to provide the best picture of the climate, scientists run tests with it. Tuning models for albedo, sea ice extent, and absolute temperatures are a few examples of the same.

Reducing the "biases formed as a result of deviations of simulations from the observed climate" is a crucial step after tuning in a climate model. These biases arise from the fact that models simplify the climate system and that the large-scale grid cells used by global models sometimes fail to capture the specifics of the local climate. Typically, this happens when regional or local simulations are used. Bias correction is often only done to model output; however in the past, it has also been used within runs of model.

As a result, GCMs are essential tools for enhancing our comprehension and ability to predict climate change. Climate modeling is used for diagnosis and prognosis.

6.4.6   Diagnostic Climate Modeling

It includes detection and attribution.

(a) *Detection*
    It is the process of confirming that the climate has changed in a specific way without identifying the cause of the change.
(b) *Attribution*
    It is the process of determining the changes that are most likely to have occurred with a certain degree of confidence.

The contribution of anthropogenic forcing to climate change in the twentieth century is an illustration of diagnostic climate modeling.

6.4.7 Prognostic Climate Modeling

It makes climate predictions for the future climate on the basis of recent or historical data (ocean structure, radiative forcing, etc.), in global warming. Seasonal/inter-annual variability, decadal prediction, and twenty-first-century scenarios are among the timescales for projection. Therefore, even though we are fully aware that the computation required for climate modeling is relatively labor-intensive, advanced algorithms and more computing power would aid in producing more accurate simulations, parameterized processes, and climate change projections.

6.4.8 Coupled Atmosphere-Ocean General Circulation Models

Coupled models are complex since they take into account various models. The coupled models' goal is to accurately and comprehensively depict how the climate system operates. For example, coupled atmosphere-ocean general circulation models (also known as "AOGCMs") can only approximate the exchange of heat and freshwater between the land surface, ocean surface, and the atmosphere.

The diversity of climate models raises a significant problem since it makes it challenging to compare the outcomes of many models because each model's methodology is challenging. In order to analyze and validate GCMs, Coupled Model Intercomparison Project (also known as "CMIP") was created as a framework for climate model experiments. These integrated atmosphere and ocean GCMs seek to enhance and unify all existing climate models. Prior to include more precise Representative Concentration Pathways, CMIP was mostly focused on modeling atmospheric CO_2 concentrations (RCPs). The findings of numerous models are loaded on a single web portal, run by the Program for Climate Model Diagnosis and Intercomparison (PCMDI), that may be freely accessed by scientists all over the world in order to rule out variations in the output using different models. CMIP is the responsibility of the Working Group on Coupled Modeling committee. It is a part of the World Climate Research Programme (WCRP) of World Meteorological Organization (WMO), Geneva. CMIP6, which consists of 21 distinct Model Intercomparison Projects, is now under way.

6.5 Greenhouse Gas Emission Scenarios

Different climate models project different values for the rise in temperature, by the next decade or next 50 to 100 years. This extreme gap between various climate models could be attributed to two factors.

Climate models are unable to factor in the effects of clouds. The clouds are composed of an important greenhouse gas called water vapor, which traps the heat, as well as exert cooling effect by blocking solar radiation from reaching the Earth's surface. Therefore, it is not clear as to which of the dual role these clouds play in altering the climate. As a result, the inclusion of these clouds causes some error in the results, which might be between 1 and 2 °C on a projection for 2100. In addition, a second factor that must be considered is the amount of greenhouse gases that would be added to the atmosphere as a result of anthropogenic inputs, in addition to the amount of greenhouse gases that are already present. Scientists refer to the behavior of greenhouse gas emissions as "emission scenarios" since the influence of clouds and greenhouse gases would change depending on various human inputs. The World Meteorological Organization (WMO) and the United Nations Environment Programme (UNEP) jointly established the Intergovernmental Panel on Climate Change (IPCC) to "assess the scientific, technical and socio-economic information relevant for the understanding of the risk of human-induced climate change." Since its founding, the IPCC has created a number of thorough Assessment Reports on the current state of knowledge regarding the causes of climate change, its probable effects, and available measures for retaliation. Additionally, it produced techniques, guidelines, technical papers, and special reports. Policymakers, scientists, and other experts frequently use these IPCC publications as standard references. The IPCC published emission scenarios in 1992 that would be used to power global circulation models and create climate change scenarios. In 1990 and 1992, the IPCC created emission scenarios for the long run. The Intergovernmental Panel on Climate Change (IPCC) released the Special Report on Emissions Scenarios (SRES) in 2000. Climate change forecasts have been made using the greenhouse gas emission scenarios discussed in the Report. A total of 40 scenarios, most of which fall into one of four main families (A1, A2, B1, B2), have been described by the IPCC. These scenarios represent a particular trajectory (evolution) of humanity, and the key hypotheses (relating to demography, agricultural practices, the spread of technology, etc.) are then translated into "food production" and "energy consumption" using models.

The first global scenarios with estimates for greenhouse gases were the 1892 scenarios. So, in 1996, the IPCC made the decision to create a new set of emission scenarios that would be more useful than the 1892 scenarios. The new scenarios offer information for assessing potential mitigation and adaptation measures as well as the climatic and environmental effects of future greenhouse gas emissions.

6.5.1 The A1 Family

The A1 family is based on the following hypothesis:

- Accelerated social and cultural exchanges.
- A world population that peaks in the middle of the twentieth century and then decreases increased economic growth.

The following three technologically focused alternative pathways for the energy system are described by the A1 scenario family: fossil intensive (AIFI), non-fossil energy sources (AIT), or balance across all sources (A1B).

According to this A1 family scenario, energy need would increase, and fossil fuel supplies would eventually run out (except coal). Additionally, in 2100, the atmospheric CO_2 concentration will be quite near to 1100 ppm.

6.5.2   The A2 Family

The A2 family is based on the following hypothesis:

- The global population is constantly growing.
- The globe is evolving in a diverse way, and economic growth is region-specific.
- The dissemination of new, efficient technology is slower and varies greatly by place.
- The underlying theme is the maintenance of local identities.

6.5.3   The B1 Family

The B1 family is based on the following hypothesis:

- The economic structures focused on information and service technology.
- The world reaches its pinnacle about mid-century and thereafter drops.
- The quick adoption of new, eco-friendly, and effective technology.
- Tackling issues related to the economy, society, and environment without taking additional climate action.

As a result, this scenario envisions the development of nuclear energy as well as significantly more oil and gas resources than are currently available. Additionally, it forecasts that in 2100, the atmospheric CO_2 concentration will be close to 450 ppm.

6.5.4   The B2 Family

The B2 family exemplifies a world where:

- By 2100, there will be 10 billion people on the planet.
- Through the use of local solutions, this family of scenarios emphasizes sustainability from the areas of economic, social, and environmental viewpoints.
- Economic growth is in the middle stages.

- It concentrates on local and regional levels and is oriented toward social fairness and environmental protection.
- The B2 family is extremely uneven in the development and transfer of efficient technologies.

In this case, the atmospheric carbon dioxide concentration reaches roughly 740 ppm.

6.6   Time-Dependent Models

The scenarios for emissions and radiative forcing mentioned include a time element, i.e., how much the climate would change and how long it would take for the change to occur. However, rather than average annual emissions, the amount of net carbon emissions that contribute to climate change is what determines its severity.

The analysis of all the aforementioned scenarios demonstrates that none of them take extreme occurrences, such as nuclear war or a widespread pandemic, into account. These scenarios also provide substantially distinct patterns for greenhouse gas emissions and concentrations over the next century, notwithstanding their rarity. An older scenario, the IS92a scenario, was employed for the 1995 IPCC report. For a given emission scenario, the various models do not differ by more than 1 to 2 °C for the predicted temperature increase in 2100.

Accordingly, various social, economic, and technological advancements have a significant impact on emission trends, and the scenarios offer crucial insight into the relationship between environmental quality and development decisions. This will help those who develop policies and make decisions on effective climate initiatives.

6.7   Representative Concentration Pathways (RCPs)

The IPCC Fifth Assessment Report (AR5) employed a set of scenarios called Representative Concentration Pathways (RCPs) to forecast future climate scenarios using climate models. The phrase "Representative Concentration Pathways" refers to the fact that these models were created to reflect potential future concentrations of greenhouse gas emissions by providing a pathway that depicts the trajectory of GHG emissions to attain a specific radiative forcing by 2100. The amount of energy that greenhouse gases and aerosols collect and hold onto in the atmosphere is measured through a process known as radiative forcing. Thus, it depends on the concentration of greenhouse gases and aerosols, changes in land cover, and total solar irradiation and can be either positive (heating) or negative (cooling). The Special

Report on Emissions Scenarios (SRES) forecasts based on socioeconomic scenarios that were used in the Third and Fourth IPCC Assessment Reports were replaced by RCP, which was accepted by the IPCC in its Fifth Assessment Report (ARS). The fundamental distinction is that RCPs fix the emission trajectory and radiative forcing that results, as opposed to the socioeconomic conditions. Thus, these RCPS can be used to examine policy choices for climate change adaptation and mitigation. Based on the amount of radiative forcing generated by greenhouse gases up until 2100, there are four routes as follows (https://www.environment.gov.au/system/files/resources/492978e6d26b-4202-ae51-Seba10c0b51a/files/wa-rep-fact-sheet.pdf).

6.7.1   RCP8.5

This trajectory predicts that by 2100:

- The radiative forcing will be greater than 8.5 W/m^2.
- In year 2100, atmospheric carbon dioxide levels would be 936 ppm (used as input in most model simulations).

According to this model, the average temperature increase from 2081 to 2100 in comparison to the baseline is 4.3 °C, with a predicted temperature increase range of 3.2 to 5.4 °C.

The average worldwide mean sea-level rise from 1986 to 2100 is 0.63 m, and the likely range is 0.45 to 0.82 m. It is predicated on doing the barest amount of effort to cut emissions.

6.7.2   RCP6

The radiative forcing is stabilized at about 6 W/m^2 after 2100 in this intermediate stabilization method:

- In year 2100, atmospheric carbon dioxide levels would be 670 ppm (used as input in most model simulations).
- According to this model, the average temperature increase between 2081 and 2100 over the baseline of 1850–1900 will be 2.8 °C, with a predicted temperature range of 2.0–3.7 °C.
- Global mean sea-level rise on average from 2081 to 2100 in comparison to 1986–2005 is 0.48 m, and the likely range is 0.30 to 0.63 m.
- Strong mitigation measures are needed, with early involvement from all emitters and active removal of atmospheric carbon dioxide after that.

6.7.3   RCP4.5

After 2100, this intermediary stabilization pathway stabilizes the radiative forcing at a value of roughly 4.5 W/m^2:

- In year 2100, atmospheric carbon dioxide levels would be 538 ppm (used as input in most model simulations).
- According to this model, the average temperature increase between 2081 and 2100 over the baseline of 1850–1900 will be 2.4 °C, with a predicted temperature range of 1.7–3.2 °C.
- The estimated range is between 0.32 and 0.63 m, with the average worldwide mean sea-level rise for the years 2081–2100 being 0.47 m in relation to 1986–2005.

6.7.4   RCP2.6

Before 2100, the radiative forcing in this pathway reaches a peak of about 3 W/m^2, after which it starts to decrease:

- In year 2100, atmospheric carbon dioxide levels would be 421 ppm (used as input in most model simulations). According to this model, the average temperature increase from 1850 to 1900 to 2081–2100 is 1.6 °C, with a likely temperature range of 0.9 to 2.3 °C.
- In comparison to 1986–2005, the average worldwide mean sea-level rise for the years 2081–2100 is 0.40 m, with a likely range of 0.26–0.55 m.
- It is additionally known as RCP3-PD. Peak and Decline is abbreviated as PD. RCP2.6 seeks to limit warming to no more than 2 °C above pre-industrial levels.

References

IPCC (2000) Summary for Policymakers Emissions Scenarios: A Special Report of IPCC Working Group III Published for the Intergovernmental Panel on Climate Change. Based on a draft prepared by: Nebojsa Nakicenovic, Ogunlade Davidson, Gerald Davis, Arnulf Grubler, Tom Kram, Emilio Lebre La Rovere, Bert Metz, Tsuneyuki Morita, William Pepper, Hugh Pitcher, Alexei Sankovski, Priyadarshi Shukla, Robert Swart, Robert Watson, Zhou Dadi. ISBN: 92-9169: 113–5

IPCC (2007) IPCC fourth assessment report: climate change 2007 (AR4). IPCC, Geneva

Moss RH, Edmonds JA, Hibbard KA, Manning MR, Rose SK, van Vuuren DP, Carter TR, Emori S, Kainuma M, Kram T, Meehl GA, JFB M, Nakicenovic N, Riahi K, Smith SJ, Stouffer RJ, Thomson AM, Weyant JP, Wilbankd TJ (2010) The next generation of scenarios for climate change research and assessment. Nature 463;747–756

Ozturk M, Hakeem KR, Faridah-Hanum I, Efe R (2015) Climate change impacts on high-altitude ecosystems. Springer Science+Business Media, XVII, New York, 695 pp

PCMDI (Program for Climate Model Diagnosis and Intercomparison) (2007) IPCC model output. Lawrence: PCMDI. Available from http://www.pemdillnl.gov/ipee/about_ipcc.php

Ramirez-Villegas J, Lau C, Köhler AK, Signer J, Jarvis A, Arnell N, Osborne T, Hooker J (2011) Climate analogues finding tomorrow's agriculture today. Paper no. 12. Cali, Colombia. CGIAR Research Program on Climate Change, Agriculture and Food Security (CCAFS). Available online at www.ceafs.cgiar.org

Chapter 7
Societal Responses to Anthropogenic Climate Change

7.1 Introduction

In the upcoming decades, climate change will have an impact on both people and ecosystems, according to a number of scientific literatures, including the IPCC's Fourth Assessment Report. In comparison to the developed nations, the developing countries are significantly impacted (Ozturk et al. 2015). According to Stern (2006), three factors contribute to the developing countries' high vulnerability: their geographic location, their heavy reliance on vulnerable industries like agriculture, and their limited capability for adaptation. The effects of climate change have not only been felt by underdeveloped countries but also by developed countries. Examples include the European heat wave of 2003 and the Katrina hurricane of 2004. Therefore, it may be inferred that the possible impact of climate change would have a major and uneven influence on people's well-being around the world and, more crucially, would act as a roadblock to sustainable development. The UNFCCC offers two policy recommendations, notably "mitigation" and "adaptation," to prevent such repercussions. Since the UNFCCC asserts that greenhouse gas emissions are the only culprit behind the climate change, it places more stress on alleviation than on adaptation (Pielke 1998). As a result, the ongoing climate change negotiations from Rio to Copenhagen placed a significant emphasis on the "mitigation" strategy. Despite several efforts, the international community has not yet been able to enact any strict mitigation measures that could prevent the potential effects in the future. The idea of "adaptation" has become the main stream in recent climate policy since there is now a dearth of robust mitigation policies and because the world has already committed to some of the potential repercussions caused by past GHG emissions in the atmosphere. Although they are not mutually exclusive and may even be complementary to one another, adaptation and mitigation are currently recognized as two crucial response strategies in climate policy (Pielke et al. 2007).

© The Author(s), under exclusive license to Springer Nature Switzerland AG 2023
M. A. Dervash et al., *Phytosequestration*, SpringerBriefs in Environmental Science, https://doi.org/10.1007/978-3-031-26921-9_7

In this chapter, adaptation and mitigation in the context of climate change shall be addressed. Additionally, we will discuss about several measures for adaptation and mitigation that have been especially applied in India. Finally, how individuals, governments, and civil society can lessen the impact of climate change shall be discussed.

7.2 Mitigation and Adaptation

While adaptation has lately acquired relevance, mitigation has a long history in climate policy. In general, the term "mitigation" refers to lowering atmospheric GHG levels so that we could reduce the chance of extreme weather events and climatic variability. "An anthropogenic intervention to reduce the sources or enhance the sinks of greenhouse gases," according to the IPCC, is what mitigation is. On the other hand, the term "adaptation" in general refers to how people alter their lives, groups, and environments to take advantage of opportunities presented by shifting social and natural systems. The adaptation of natural and human systems to existing or anticipated climatic stimuli, which can lessen the detrimental effects and maximize the positive, is how climate change is defined in the literature (UNFCC 1992).

The setting and scale are two factors that specifically affect adaptation. At the farmer level, switching to new hybrid seeds might be one type of adaptation. At the farm level, insurance and diversification might be necessary. At the regional level, it might be related to the number of farms participating in a compensation program. At the global level, it might involve a change in the pattern of international food trade (Kumar 2009). Irrigation may be a suitable adaptation method for dry land agriculture depending on scale, but it depends on groundwater supply over the long term (Kumar 2009). The literature has asked important issues such as "adaptation to what," "who or what adapts," "how does adaptation occur," "what and how resources are used," and "how good is the adaptation" in an effort to methodically develop a framework that defines the concept. Although the diversity of adaptation implies that there is no single approach for evaluating, planning, and putting into action adaptation measures, it can be reviewed in light of the various types of adaptation that are already in place, including purposefulness, timing, temporal scope, spatial scope, function/effects, form, and performance.

As already stated, there are significant non-linear consequences of climate change on the health of human society. Many developing countries have already been affected by weather-related severe events, such as floods, droughts, heat waves, and tropical cyclones, which are occurring more recurrently or intensely than in the past. Generally speaking, it has an impact on a variety of industries, including freshwater resources and their management; food, fiber, and forest products; coastal systems and low-lying areas; health; etc. Future climate variability and change will have enormous effects on the environment, production systems, and way of life. It is imperative to note that developing countries have more obligations than developed countries (Stern 2006; Mendelsohn et al. 2006). According to a scientific research, "if we don't act, the overall damage cost will be comparable to at least 5 percent of GDP now and forever, and if a wider range of risks and repercussions is taken into account, the estimates of damage might reach to 20 percent of GDP or more" (Stern 2006).

With the aforementioned context in mind, it has compelled the international community to think about preventative measures to lessen hazards related to climate change. The IPCC's AR4 found that several consequences can be avoided, miti-gated, or deferred by pursuing mitigation actions and that a range of adaptation techniques could lessen the risks associated with climate change (Dutt and Gaioli 2008). Additionally, mitigation and adaptation are seen as complementary in terms of climate policy as the former can work to minimize the impact component (sensi-tivity), while the latter can work to reduce the likelihood component of a risk calcu-lation (exposure) (Yohe and Strzepek 2007).

In general, mitigation involves bringing atmospheric concentrations of GHGs to a predetermined level, and it specifically depends on how much potential impact we are willing to accept. William Nordhaus, for instance, suggested that the ideal rate of emission reduction is 10–15%, while Stern suggested a reduction of 30–70 over the next two decades (Nordhaus 2007). GHG emissions have risen by 1.6% per year on average over the past three decades, and under the existing climate policy frame-work, it is anticipated that they will continue to rise (Rogner et al. 2007). The demand for and supply of energy are expected to rise globally despite ongoing improvements in energy intensities, particularly in developing countries where industrialization has lately become a priority. If there isn't a significant shift in energy policy, the energy mix used to power the world economy in the period between 2025 and 2030 will virtually stay the same. Fossil fuels will account for more than 80% of the energy supply, which will have an impact on GHG emissions (Rogner et al. 2007). As a result, it will be too expensive and difficult to cut the GHG emissions that the developed countries are already experiencing. This is because it would place developing countries like China and India in a carbon lock-in effect. This is the reason why, in addition to the industrialized countries, emerging developing countries must also reduce their GHG emissions.

Additionally, under the present climate policy, "adaptation is recognized as a beneficial instrument combined with mitigation. There are six factors in this situa-tion that motivate policymakers and the vulnerable communities in the region to concentrate on adaptation," including (McCarthy et al. 2001; Stern 2006; Pielke et al. 2007):

 (i) Because of past emissions of GHGs into the atmosphere, some of the irrevers-ible effects of climate change cannot be completely averted.
 (ii) In comparison to forced and reactive adaptation, anticipatory and preventative action is more efficient, less expensive, and sustainable.
(iii) Climate change could have more severe effects than currently thought, includ-ing many irreversible effects (e.g., severe hurricane, melting of Greenland ice sheets fully, etc.). On the other hand, because several million people are already living in poverty, the climatic risk will have a non-linear and more difficult influence on their standard of living.

(iv) Improved adaptation to climate variability and extreme atmospheric occurrences, along with the abolition of unfavorable policies and behaviors, can have an immediate positive impact.
(v) Threats and possibilities are brought about by climate change, and benefits may accrue in the future.
(vi) The poor developing countries are pressing for increased international "adaptation" as a kind of international responses to help them become more resilient.

The IPCC's Third Assessment Report made clear that both adaptation and mitigation measures affect how severe the effects of climate change will be. While mitigation would be considered indirect damage protection, adaptation might be interpreted as direct harm prevention (Klein et al. 2005). Since we have already agreed to some of the climate effects brought on by past atmospheric GHG emissions, adaptation is currently necessary. Without mitigation, climate change is predicted to reach a point where some natural systems cannot adapt and most human systems must bear very large social and economic costs. As a result, there is a chance to incorporate both adaptation and mitigation into more comprehensive development strategies and policies. Recently, policymakers have shown an interest in investigating how adaptation and mitigation are related to one another. They are, however, confronted with a wide range of conceptual problems, including how much adaptation and mitigation would be ideal, when and how to combine them for substitution, complementarily when and where to invest in adaptation and mitigation, how their costs and efficacy change over time, how the two responses interact, how development pathways affect them, etc. There are four interrelationships between mitigation and adaptation in the context of climate policy:

(i) Adaptation measures with implications for mitigation.
(ii) Actions taken for mitigation have adaptation-related repercussions.
(iii) Synergistic decisions between adaptation and mitigation.
(iv) Processes with effects on both mitigation and adaptation.

Additionally, there is a variety in both adaptation and mitigation. At various decision-making levels, adaptation and mitigation are related in a variety of ways. If mitigation measures remove market distortions and failures, as well as perverse subsidies that hinder decision-makers from considering the full social costs of the available options, they can promote adaptive capacity. In actuality, the actors and budgets involved are different, despite the fact that at a huge scale, mitigation expenditures seem to redirect social or private resources and lower the amount of money available for adaptation. Although direct trade-offs are uncommon, both alternatives alter relative pricing, which can result in minor changes to consumption and investment patterns and, ultimately, alter the course of the affected economy's development. The effects of adaptation on mitigation can be both beneficial and detrimental. For instance, afforestation as a component of a regional adaptation strategy helps with mitigation. On the other hand, adaptive measures that necessitate greater energy use from carbon-emitting sources (such as interior cooling) will harm mitigation efforts.

7.3 Climate Change Debates

The United Nations Framework Convention on Climate Change (UNFCCC) laid the foundation of a treaty of an international repute in 1992. Popularly known as United Nations Conference on Environment and Development (Earth Summit), it was held from 3 to 14 June 1992 and entered into an action on 21 March 1994. Conference of the Parties (COP) is basically a governing council of an international convention which serves as the formal meeting of the countries (which are parties to UNFCCC) to monitor the overall performance of climate change mitigation scenarios in context of anthropogenic interventions. The meetings are meant to ascertain legally bound commitments for developed nations to curtail their greenhouse gas emissions. The final conclusions of these meetings must be taken up on consensus.

Since the Rio Earth Summit (1992), there has been a great deal of discussions and advancements related to climate change. However, we would discuss about the Copenhagen Summit and the Kyoto Protocol, two significant events (Table 7.1).

7.3.1 Kyoto Protocol

It is a global agreement to curtail greenhouse gas emissions supported by the United Nations. The United Nations Framework Convention on Climate Change, which was ratified by almost all of the countries present at the Rio Earth Summit (1992), gave rise to this protocol. GHG concentrations are to be maintained "at a level that would preclude dangerous anthropogenic interaction with the climate system," according to the Framework. The Protocol was first ratified on 11 December 1997 in Kyoto, Japan, and went into effect on 16 February 2005. One hundred eighty-seven nations had signed and approved the treaty till November 2009. The United States, which is a signatory to UNFCCC and was in charge of 36.1% of 1990's emission levels, is the most prominent non-member of the Protocol.

According to the Protocol, 37 developed nations (referred to as "Annex I countries") agree to trim down four greenhouse gases (GHGs), including CO_2, CH_4, N_2O, and SF_6, as well as two sets of gases created by them, HFCs and PFCs. The nations listed in Annex I have promised to cut their total emissions of greenhouse gases by 5.2% from 1990 levels. The Montreal Protocol on Substances that Deplete the Ozone Layer (1987) regulates industrial gases, such as chlorofluorocarbons or CFCs, but does not address emissions from international commerce and aviation. The benchmark 1990 emission levels were determined by the "global warming potential" numbers derived for the IPCC Second Assessment Report and approved by the Conference of the Parties of the UNFCCC (decision 2/CP.3). When calculating total sources and sinks, these numbers are utilized to convert different greenhouse gas emissions into comparable CO_2 equivalents.

Table 7.1 The list of COP meetings

COP	Date	Host country
COP 1	28-03-1995 to 07-04-1995	Berlin, Germany
COP 2	08-07-1996 to 19-07-1996	Geneva, Switzerland
COP 3	1-12-1997 to 10-12-1997	Kyoto, Japan
COP 4	2-11-1998 to 13-11-1998	Buenos Aires, Argentina
COP 5	25-10-1999 to 5-11-1999	Bonn, Germany
COP 6	13-11-2000 to 25-11-2000	The Hague, Netherlands
COP 6 Bis	17-07-2001 to 27-07-2001	Bonn, Germany
COP 7	29-10-2001 to 10-11-2001	Marrakech, Morocco
COP 8	23-10-2002 to 1-11-2002	New Delhi, India
COP 9	01-12-2003 to 12-12-2003	Milan, Italy
COP 10	06-12-2004 to 17-12-2004	Buenos Aires, Argentina
COP 11/CMP 1 (CMP: *Meeting of the Parties to the Kyoto Protocol*)	28-11-2005 to 9-12-2005	Montreal, Canada
COP 12/CMP 2	06-11-2006 to 17-11-2006	Nairobi, Kenya
COP 13/CMP 3	03-12-2007 to 14-12-2007	Bali, Indonesia
COP 14/CMP4	01-12-2008 to 12-12-2008	Poznan, Poland
COP 15/CMP 5	07-12-2008 to 18-12-2009	Copenhagen, Denmark
COP 16/CMP 6	28-11-2010 to 10-12-2010	Cancun, Mexico
COP 17/CMP 7	28-11-2011to 09-12-2011	Durban, South Africa
COP 18/CMP 8	26-11-2012 to 07-12-2012	Doha, Qatar
COP 19/CMP 9	11-11-2013 to 23-11-2013	Warsaw, Poland
COP 20/CMP 10	01-12-2014 to12-12-2014	Lima, Peru
COP 21/CMP 11	30-11-2015 to 12-12-2015	Paris, France
COP 22/ CMP 12/CMA 1(CMA: *Meeting of the Parties to the Paris Agreement*)	07-11-2016 to 18-11-2016	Marrakech, Morocco
COP 23/CMP 13/CMA 1-2	06-11-2017 to17-11-2017	Bonn, Germany
COP 24/CMP 14/CMA 1-3	03-12-2018 to14-12-2018	Katowice, Poland
COP 25/CMP 15/CMA 2	11-11-2019 to 22-11-2019	Madrid, Spain
COP 26/CMP 16/CMA 3	31-10-2021 to 12-11-2021 (*Due to Covid-19 Pandemic, COP 26 was not undertaken in 2020*)	Glasgow, United Kingdom
COP 27/CMP 17/CMA 4	06-11-2022 to 18-11-2022	Sharm El-Sheikh, Egypt

The main query is whether Kyoto will have any impact. However, it seems that just one or two of the major economies will achieve their Kyoto targets by 2012. Even if they succeeded, it would only make a very little difference in the global output of GHGs. Because atmospheric CO_2 has a century-long half-life, the world has already committed to a significant quantity of greenhouse warming. Climate scientists predict that the planet would rise by at least 0.5 °C more if all fuel-burning devices were turned off tomorrow. The basic line is that until technological and societal advancements make it possible for us to significantly reduce our emissions, we will not be able to control greenhouse gas emissions. To remove massive amounts of carbon or both, we need to develop some safe techniques.

7.3.2   Copenhagen Summit

The Copenhagen Summit, also known as the United Nations Climate Change Conference (2009), took place in Copenhagen, Denmark, from December 7–18. A framework for reducing climate change after 2012 was supposed to be agreed upon, under the Bali Road Map. The US administration deemed the Copenhagen Accord to be a "meaningful accord, which was drafted on 18th December by the United States, China, India, Brazil, and South Africa." In a discussion among all the participating nations the next day, it was "took note of," but not "accepted," and it was not approved by a unanimous vote. The statement acknowledged that one of the biggest challenges of our time is climate change and that steps should be done to keep any temperature increases less than 2 °C. Countries most vulnerable to climate consequences did not get the agreement they desired since there is no hard target for limiting global temperature escalation, no commitment to a binding treaty, and no target year for depreciating the increased emissions.

Delegates voted in favor of a proposal to "take note of the Copenhagen Accord of December 18th, 2009" early on December 19. This was brought about by the opposition of nations including Bolivia, Venezuela, Sudan, and Tuvalu who voiced their opposition to the goals and procedures used to reach the Copenhagen Accord. The agreement acknowledges the scientific rationale for limiting temperature increases to 2 °C, but it does not include pledges to reduce emissions, which would be required to accomplish that goal. As part of the deal, the developing world were supposed to receive US$ 30 billion over the three subsequent years, up to US$ 100 billion annually by 2020, to aid in their efforts to adapt to climate change. Prior plans that called for a 1.5 °C maximum increase in temperature and an 80% reduction in CO_2 emissions by 2050 were abandoned. The Accord also supported "REDD," a program in which wealthier nations pay developing nations to reduce emissions and environmental deterioration.

Despite widespread belief that the Copenhagen meeting would result in a binding agreement, negotiations broke down, and the so-called Copenhagen Accord was never able to become a formal agreement. The Copenhagen Accord paved the ground for additional discussions to take place at the 2010 UN Climate Change Summit in Mexico and required nations to submit emission targets by the end of January 2010. Sixty-seven countries had registered their targets by the beginning of February. The Copenhagen Accord cannot replace negotiations under the UNFCCC, according to nations like India and the Association of Island States. The future of the UN's participation in global climate agreements, according to other commentators, is now in doubt. The Copenhagen Accord, according to Indian journalist Praful Bidwai, "is an illegal, poorly thought-out, collusive deal between a handful of countries that are some of the world's greatest present and future emitters." He places the blame on both developed and a few developing nations, including India. He contends that elites are seeking to maintain high levels of their emissions by indirectly increasing the disproportionate burden on world's developing countries.

7.3.3   COP 27: Sharm El-Sheikh, Egypt, from 6 to 18 November 2022

The major highlights of COP 27 are as follows: the rich countries should come forward to support poor countries in terms of adaptation financing through collaborations, Bill & Melinda Gates Foundation consented to donate $1.5 billion in technology investment to curb climate change, United States made a statement to curtail their carbon footprints by 2030, and Egypt proclaimed to increase the sustainability as well as resilience in Agriculture Sector through Food and Agriculture for Sustainable Transformation (FAST) and also urged to ascertain Initiative on Climate Action and Nutrition (I-CAN). Egypt also proposed to focus on women's inclusion in climate action and adaptation as well mitigation strategies. In a nutshell, all the representative delegates urged upon strengthening the adaptation and resilience strategies to the changing climate across the world through international collaborators and through adoption of climate smart technologies for sustainable development.

The UN Climate Change High-Level Champions took stock of the contribution of non-state actors at COP 27 in *Sharm el-Sheikh, Egypt, on November 17*. The champions put forward a wide range of actions, announcements, and events on mitigation, adaptation finance, and equity, with a clear focus on implementation, including the launch of the African Cities Water Adaptation Fund, an African-led insurance commitment to cover up to 14 billion dollars in climate losses. A new 5-year work program was launched to promote climate technology solutions in developing countries. TEC and CTCN submitted a joint work program to accelerate the deployment of transformative climate technologies, urgently required to tackle climate change; an important opportunity to step up rapidly efforts to deploy technology to address through mitigation and adaptation, covering the period from 2023 to 2027. Specific joint activities are expected to be implemented by the TEC and CTCN, including technology roadmaps, work on gender and technology, technology and NDCs and digitalization, and common areas of work, including national systems of innovation, industry, and the water-energy-food nexus. There is a need for a rapid scaling up and effective transfer of climate technologies to limit the global average temperature rise to 1.5 °C and build resilience to climate change. Better Technology Mechanism is required to unlock appropriate climate technologies globally. IPCC AR6 and the contributions from WG III on Mitigation of Climate Change point to the technology as an enabler to accelerate mitigation and to drive effective adaptation solutions. It stresses upon an averment that limiting global warming will require significant transitions in energy, in cities, particularly to the key areas with high potential for emission reductions like cities/urban areas, industry, agriculture, forestry, and land use change. This joint work program is a significant milestone for the Technology Mechanism guided by science focusing on high-potential sectors and high-impact actions. The polluters are committed to ensuring good cooperation, effective to accelerate action at the scale and pace needed through research, development, demonstration, and deployment of climate technologies. According to the European Commission, technology transfer is important for the success of the global green transition in energy, infrastructure, mobility, or food. There is a need for sharing the knowledge to help build up industrial capacity and support developing countries to tackle the climate crisis. The new program opens new possibilities for innovation and targeted action. Germany has announced a voluntary contribution of EUR 1.5 million to CTCN and EUR 500.000 to TEC in 2022. Accelerated and fast deployment of climate technologies is a highly important building block in the set of solutions for transformational change necessary to reach earlier targets. The new joint work program marks an important milestone in the further development of the Technology Mechanism. This new joint program will help globally collaborate. Japan has committed to continue its current level of funding, and CTCN is strengthening collaboration with Korea through the recent establishment of its Partnership and Liaison Office in Songdo, Republic of Korea.

The major highlights of COP 27 are as follows: the rich countries should come forward to support poor Bill & Melinda Gates Foundation consented to donate $1.5 billion in technology investment to curb climate change, the United States made a statement to curtail their carbon footprints by 2030, and Egypt proclaimed to increase in the sustainability as well as resilience in Agriculture Sector through Food and Agriculture for Sustainable Transformation (FAST) and also urged to ascertain Initiative on Climate Action and Nutrition (I-CAN). Egypt also proposed to focus on women's inclusion in climate action and adaptation as well mitigation strategies. In a nutshell, all the representative delegates urged upon strengthening the adaptation and resilience strategies to the changing climate across the world through international collaborators and through adoption of climate smart technologies for sustainable development. The details are outlined as follows: *UN Climate Change meeting on 15 November 2022* has been the first high-level roundtable on *pre-2030* at COP 27. This was meant for a collective call to urgently ramp up climate action and support, to set the global direction on mitigation ambition and implementation that should be evaluated. As per the report, implementation of current pledges by national governments will increase emissions by 10.6% by 2030. It will put our globe on track for a 2.5 °C warmer world by the end of the century. There is a bend in the curve of greenhouse gas emissions downward at the global level, but the efforts are insufficient to limit global temperature rise to 1.5 °C. According to IPCC, greenhouse gas emissions must peak before 2025 at the latest and decline 43% by 2030 to limit global warming to 1.5 °C. We have to announce it to the policymakers, markets, and individuals from COP 27 that this is happening, and we must reduce emissions faster, catalyze impactful action, and secure assurances from key countries that they will take immediate action to raise ambition and keep us on the path toward 1.5 °C. The conference decision on the mitigation work program must stress that all parties support this statement. It must reflect the level of urgency, the gravity of the threats we are facing, and the shortness of the time we have remaining to avoid the devastating consequences of runaway climate change. As per IPCC's Working Group III, there is a sense of hope that there are *solutions in all sectors* that can be deployed immediately to close the gap by 2030. The mitigation solutions have the potential to halve emissions by 2030, costing less than 100 dollars/ton of CO_2 equivalent. The costs of wind and solar have dropped dramatically, and there is a huge potential of emission reductions in agriculture, land use, and forestry, including better soil management and ecosystem restoration. A collective call was to urgently ramp up ambition, with many developing countries stressing the need for support and sustained financial flows. Most ministers agreed the 1.5 °C temperature limit is a "red line" that cannot be crossed.

7.4   Role of Individual, State, and Civil Society for Sustainable Adaptation

Today, it is best to understand the nature of global governance on a variety of topics, including climate change, as an amalgamation of what the state, civil society, and markets do or do not. This is not meant to imply that they constantly or even primarily collaborate or have the same priorities. It means that they do so more frequently than they previously did, including in terms of reducing global climate change (Sathaye et al. 2007). Generally speaking, we think that only the government can act in the public interest, while business and citizens are primarily interested in their own interests. Others, however, believe that the market and competition provide the best results, both public and private, since they believe that all actors are motivated by self-interest. This point of view contends that while the state continues to play a role in establishing rules and monitoring performance, civil society, consumers, and industry face greater responsibility and share the risks. It is commonly agreed that accountability for the environment and sustainability have expanded into a much bigger enterprise, even though the duties, responsibilities, and authority given to each player are still highly contested. Governments are no longer solely responsible for it; instead, the state, the corporate sector, and civil society are all involved (Sathaye et al. 2007; Ozturk et al. 2015).

7.4.1   Role of Individuals to Slow Down Climate Change

Although there is a huge concern with climate change, individuals can help minimize greenhouse gas emissions and thereby the damaging consequences of climate change by adopting conscious practices like:

- Share your knowledge of climate change.
- Invest in more energy-efficient home appliances.
- Compact fluorescent bulbs can replace all incandescent lighting and last four times as long while using only one-fourth of the energy.
- Put an emphasis on green building designs.
- Drive more fuel efficiently.
- Avoid leaving the engine running on a prolonged basis.
- As much as possible, stay away from motorized vehicles (France and Italy have "No Car Days" and only allow parking in cities on alternating days for license plates with odd and even numbers).
- When not in use, turn off all lights, televisions, fans, air conditioners, computers, and other electrical devices.
- Plantation drives at a broader scale.
- Buy recycled goods as often as you can, and recycle all cans, bottles, and plastic bags.
- Produce as little garbage or waste as you can, as garbage in landfills releases a lot of methane and when it burns, it releases CO_2.

7.4.2   Role of State

The shift from government to governance takes into account the evolving political patterns in both developed and emerging nations (Tosun and Howlett 2022). Importantly, the pace and extent of institutional reforms by the government and market activity—for example, the potency and engagement of civil society organizations and the organization of the financial and capital markets, corporate governance, and corporate social responsibility—depend on the specific state (Sathaye et al. 2007). These reforms are particularly reliant on a nation's pre-existing institutions, local reform politics, and internal interests that are refractory to change. The choice of potential national growth trajectories in major input sectors will be significantly impacted by the restructuring of governance institutions. The political climate and regulatory policy framework of a nation or area, as well as the degree of public expectation that their governments, will take a strong or weak lead in pursuing policy solutions, all have an impact on the types of policies that governments desire and are able to practice.

National policy styles or political cultures are identified and explained by a sizable body of political theory. The underlying premise is that each country has a tendency to approach issues in a particular way, regardless of how distinctive or unique a situation may be; this is known as a national "way of doing things." The salient characteristics of prevalent "policy styles" in different nations and areas worldwide are emphasized. National policy style was defined by Richardson et al. (1982) as originating from the interaction of two components. Using a fundamental typology of styles, countries are classified according to whether national decision-making is anticipatory or reactive and and whether the political climate is consensus-based or an imposition (Tosun and Howlett 2022).

The institutional ability of governments to implement the policy instrument on the ground is a crucial, but frequently ignored, factor in the choice of policy instruments (Rayner 1993; Tosun and Howlett 2022). This frequently depends on what nations with severely limited resources believe they can afford. These criteria consist of:

- A strong institutional framework for enforcing regulations.
- An economy that exhibits certain traits of free market economic models and is therefore likely to respond well to fiscal policy tools.
- A highly developed informational sector and mass media infrastructure for public opinion formation, advertising, and education.
- A significant annual R&D expenditure for both the public and private sectors that is used to launch pilot programs and reduce uncertainty (O'Riordan et al. 1998).

So far these nearly optimal conditions for traditional policy instruments are lacking. Effectiveness barriers for policymakers are likely to exist (Tosun and Howlett 2022). For instance, the meticulously developed forest protection laws in Brazil and Indonesia (Petrich 1993) could be implemented to ensure forest preservation and carbon management. To ensure that those laws are effective, however, neither nation is able to devote enough resources to monitoring and enforcement. The urge to use natural resources to generate foreign cash is a major contributor to the paucity of program resources in much of the developing nations. This puts more pressure on conversion of forest area to habitation and raises population's demand for energy. Legislative actions consequently frequently seem to "leave more imprints on paper than on the landscape" (Rayner and Richards 1994).

Poor infrastructure is common in less developed nations, which is made worse by a lack of human, financial, and technological resources. Additionally, these nations are probably going to concentrate on more fundamental nation-building and economic growth issues. The economic conditions of less industrialized nations also offer chances to implement emission reduction measures and sustainable development goals at a cheaper cost than in industrialized nations. The concepts of adaptive and mitigating capacity put forth in the IPCCTAR seem to support the view that the capacity to create and implement strategies for responding to climate change is fundamentally the same as that needed to create and execute policies across a wide range of sectors. They basically share the same concepts as sustainable development. According to O'Riordan et al. (1998), "the more that climate change issues are routinely integrated into the planning perspective at the appropriate level of implementation, the national and local government, the enterprise, the community, the more likely they are to achieve desired goals." It is challenging to execute significant climate policies as a whole. However, it is unlikely to be successful to merely tack climate change onto an already established political agenda.

7.4.3   Role of Civil Society

The term "civil society" refers to the setting for voluntary group action centered on common goals, objectives, and ideals. Commonly, civil society embraces a variety of places, people, and institutional structures that range in formality, autonomy, and authority. Registered charities, non-governmental organizations (NGOss), community groups, women's organizations, faith-based organizations, professional associations, trade unions, self-help groups, social movements, business associations, coalitions, and advocacy groups are just a few examples of the organizations that frequently make up civil societies (Sathaye et al. 2007).

Most people are aware with the idea of civil society advocating for environmental protection and legislation to reduce climate change. There are many instances of civil society groups and movements attempting to influence policy change at the international, national, and even local levels. Different interest groups within civil society may have different reform goals. But what unites them all is the legitimate role that civil society plays in outlining and pursuing its ideals of change through a variety of channels, including public advocacy, voter education, lobbying policy-makers, research, and demonstrations in public. Given the complexity of the problem, civil society encompasses not only non-governmental organizations (NGOs) but also academic and other non-governmental research institutions, business organizations, and, more broadly, "epistemic or knowledge communities that work on better understanding of the climate change problematic."

Some climatologists have suggested that civil society has played a crucial role in bringing the issue of global climate change into the political spotlight and steadfastly promoting its significance. Governments have finally started to act on these requests from the public sector for comprehensive environmental protection and global climate change mitigation measures. The significance of non-governmental and civil society actors in advancing global climate change mitigation is particularly highlighted by studies on the negotiation processes of global climate change policy. Actually, the IPCC assessment process is a volunteer knowledge community that aims to organize the current body of information on climate change for decision-makers. It serves as an illustration of how civil society and, in particular "epistemic or knowledge communities," can directly contribute to or "pull" the discussion on international climate policy. Additionally, NGOs and the knowledge communities have played a crucial role in meeting the demands of national and sub-national climate policy. Two examples of how civil society has created forums and spaces for discourse by different actors and not just civil society actors to interact and advance the conversation on the direction that climate change mitigation and sustainable development policy should take are the Pew Climate Initiative and the Millennium Ecosystem Assessment. These civil society forums are becoming increasingly aware of the need to increase involvement from other institutional spheres of society.

References

Dutt G, Gaioli F (2008) Negotiations and agreements on climate change at Bali. Econ Polit Wkly 43:11–13

Klein RJT, Schipper ELF, Dessai S (2005) Integrating mitigation and adaptation into climate and development policy: three research questions. Environ Sci Policy 8:579–588

Kumar KSK (2009) Climate sensitivity of Indian agriculture: do spatial effects matter? Working Paper no. 45-09, South Asian Network for Development and Environmental Economics (SANDEE), Kathmandu, Nepal

McCarthy JJ, Canziani OF, Leary NA, Dokken DJ, White KS (2001) WG II: climate change 2001: impacts. Adaptation & vulnerability. Cambridge University Press, p 1000

Mendelsohn R, Dinar A, Williams L (2006) The distributional impacts of climate change on rich and poor countries. Environ Dev Econ 11:159–178

Nordhaus WD (2007) A review of the Stern review on the economics of climate change. J Econ Lit 45:686–702

O'Riordan T, Cooper CL, Jordan A, Rayner S, Richards K, Runci P, Yoffe S (1998) Institutional frameworks for political action. Human choice and climate change. Battelle Press, Ohio, pp 345–370

Ozturk M, Hakeem KR, Faridah-Hanum I, Efe R (2015) Climate change impacts on high-altitude ecosystems. Springer Science+Business Media, XVII, New York, 695 pp

Petrich CH (1993) Indonesia and global climate change negotiations: potential opportunities and constraints for participation, leadership, and commitment. Glob Environ Chang 3(1):53–77

Pielke RA (1998) Rethinking the role of adaptation in climate policy. Glob Environ Chang 8(2):159–170

Pielke RA, Gwyn P, Steve R, Daniel S (2007) Lifting the taboo on adaptation. Nature 455(8):597–598

Rayner S (1993) National case studies of institutional capabilities to implement greenhouse gas reductions. Glob Environ Chang 3(Special issue):7–11

Rayner S, Richards KR (1994) I think that I shall never see … a lovely forestry policy. Land use programs for conservation of forests. In Workshop of IPCC WGIII, Tsukuba

Richardson J, Gustafsson G, Jordan G (1982) The concept of policy style

Rogner HH, Zhou D, Bradley R, Crabbé P, Edenhofer O, Hare B, Kuijpers L, Yamaguchi M (2007) Introduction. In: Metz B, Davidson OR, Bosch PR, Dave R, Meyer LA (eds) Climate change 2007: mitigation. Contribution of Working Group III tothe Fourth Assessment Report of the Intergovernmental Panel on Climate Change. Cambridge University Press, Cambridge/New York

Sathaye J, Najam A, Cocklin C, Heller T, Lecocq F, Llanes J, Pan J, Rayner S, Robinsin J, Schaeffer R, Sokona Y, Swart RJ, Winkler H (2007) Sustainable development and mitigation. In: Metz B, Davidson OR, Bosch PR, Dave R, Meyer LA (eds) Climate change 2007: mitigation. Contribution of Working Group III to the FourthAssessment Report of the Intergovernmental Panel on Climate Change. Cambridge University Press, Cambridge/New York

Stern N (2006) Stern's review on Economics of climate change. Cambridge University Press, Cambridge/New York

Tosun J, Howlett M (2022) Analyzing national policy styles empirically using the sustainable governance indicators (SGI): insights into long-term patterns of policy-making. Eur Policy Anal 8:160. https://doi.org/10.1002/epa2.1142

United Nations Framework Convention On Climate Change (UNFCC) (1992) UN General Assembly, United Nations Framework Convention on Climate Change : resolution / adopted by the General Assembly, 20 January 1994, A/RES/48/189. Available at: https://www.refworld.org/docid/3b00f2770.html. Accessed 21 Sept 2022

Yohe G, Strzepek K (2007) Adaptation and mitigation as complementary tools for reducing the risk of climate impacts. Mitig Adapt Strat Glob Change 12:727–739. https://doi.org/10.1007/s11027-007-9096-3

Chapter 8
Monitoring of Nutrient Pollution in Water

8.1 Introduction

Organisms depend upon nutrients for survival, growth, and reproduction. The autotrophs rely upon inorganic nutrients (carbon dioxide, nitrate, and phosphate), the visible solar radiation (400–700 nm), and through photosynthesis formulate the organic molecules that make cell organelles and facilitate growth, survival, and reproduction. In contrast to autotrophs (algae and macrophytes), heterotrophs require ready-made organic molecules for meeting nutritional and energy requirements. Fungi and the majority of bacteria (except cyanobacteria) are heterotrophs that require wide diversity of inorganic and organic compounds for nutrition. Thus, the basic supply of inorganic nutrients and sunlight regulate how rapidly organisms can produce biomass in an ecosystem. Inorganic nutrients are produced naturally in ecosystems through the weathering of rocks; the biological nitrogen fixation, which transforms atmospheric nitrogen (N_2) into ammonia (NH_3); and the microbial and fungal decomposition of dead organisms. Runoff and subterranean groundwater flows are two ways that nutrients are brought to rivers and lakes. Nutrient concentrations that occur naturally vary among rivers in different geological and climatic locations due to differences in rock types, terrestrial flora that sequesters nutrients, and precipitation (Ozturk et al. 1996; Ozturk and Altay 2018; Poikane et al. 2021). The majority of ecosystems that do not experience human-caused nutrient pollution have very low nutrient concentrations because nutrient production activities are often very low in comparison to demand in ecosystems. Among all the nutrients in aquatic environments, the macronutrients nitrogen and phosphorus are crucial because they are typically in lower concentrations than the others. The rate at which algae and plants can develop is constrained when they are scarce, and aquatic plants flourish when these nutrients are in abundance. Algae and plants utilize phosphorus and nitrogen as phosphate, nitrate, and ammonia. In general, nitrogen tends to limit terrestrial and marine ecosystems more than phosphorus, while phosphorus tends to

M. A. Dervash et al., *Phytosequestration*, SpringerBriefs in Environmental Science, https://doi.org/10.1007/978-3-031-26921-9_8

Fig. 8.1 Aquatic monitoring, an essential tool for determination of nutrient pollution

limit freshwater habitats more than nitrogen. The majority of nutrient contamination is caused by over-fertilization of terrestrial environments (especially agricultural enterprise) and the nutrient-rich wastes generated by humans and other
animals. The enrichment of water bodies with these nutrients due to an excess of
these nutrients is harmful to the aquatic system's natural equilibrium. Therefore, it
is crucial to ascertain the nutrient composition of every water body (Fig. 8.1).

8.2   Nutrients

The most significant nutrients for plant growth are phosphorus and nitrogen, which
are adsorbed and actively exchanged between sediment and water. Sediment-bound
nutrients form a reserve pool that, under certain circumstances, can be released back
into the waters above, amplifying the effects of nutrient enrichment (eutrophication)
(Qin et al. 2006).

8.3   Nitrogen Compounds

Since nitrogen is a crucial component of proteins, including genetic material, it is
necessary for all life forms. Inorganic nitrogen is transformed into organic forms by
plants and microorganisms. Inorganic nitrogen (NO^{3-} and NO^{2-}), the ammonium
ion (NH^{4+}), and molecular nitrogen (N_2) are among the various oxidation states that
can be found in the environment. It changes in the environment through biological

and non-biological processes as part of the nitrogen cycle. Phase changes including volatilization, sorption, and sedimentation are key non-biological processes. The biological processes include the following: (a) plants and microorganisms assimilate inorganic forms of nitrogen (ammonia and nitrate) to produce organic nitrogen, such as amino acids; (b) microorganisms reduce nitrogen gas to ammonia and organic nitrogen; (c) complex heterotrophic conversions between different organisms; (d) oxidation of ammonia to nitrate and nitrite (nitrification); (e) ammonification; and (f) denitrification under anoxic conditions. It is highly recommended that all nitrogen species be stated in moles per liter or as mg l^{-1} of nitrogen (e.g., NO^{3-}-N, NH^{4-}-N), rather than as mg l^{-1} of NO^{3-} or NH^{4+}, to better comprehend the nitrogen cycle.

8.3.1 Ammonia

Natural sources of ammonia include the breakdown of nitrogenous organic and inorganic substances in soil and water, biota excretion, microbial reduction of nitrogen gas in water, and gaseous exchange with the atmosphere. It is also released into water bodies as a byproduct of various industrial processes, such as the production of pulp and paper using ammonia, as well as from municipal waste. High ammonia (NH_3) concentrations are harmful to the ecological balance of water bodies at specific pH values because they are hazardous to aquatic life. Unionized ammonia and the ammonium ion coexist in equilibrium in aqueous solutions. These two types add up to total ammonia (Hedayati and Sargolzaei 2013). Additionally, ammonia can adsorb to colloidal particles, suspended sediments, and bed sediments in the form of complexes with various metal ions. Additionally, it might be transferred from the water below to the sediments. Temperature, pH, and total ammonia concentration all affect the concentration of unionized ammonia.

Small levels of ammonia and ammonia compounds, typically 0.1 mg l^{-1} as nitrogen, are present in clean waters. The average amount of total ammonia found in surface waters is less than 0.2 mg l^{-1} N, but it can occasionally rise to 2–3 mg l^{-1} N. Higher amounts might be a sign of organic contamination from runoff from agricultural fields, sewage from households, or industrial waste. Therefore, ammonia is a helpful indicator of organic contamination. The decomposition of aquatic organisms, particularly phytoplankton and bacteria in nutrient-rich waters, results in natural seasonal oscillations as well. The bottom waters of lakes that have anoxic conditions may also have high levels of ammonia. Ammonia samples should be analyzed as soon as possible, ideally within 24 h. If this isn't practicable, the sample can be kept by being deep frozen or by adding 0.8 ml of sulfuric acid (H_2SO_4) to each liter of sample and storing it at 4 °C afterward. Any acid employed as a preservative should be neutralized before analysis. The measurement of ammonia ions can be done using a variety of techniques. The simplest are colorimetric procedures employing Nessler's reagent or the phenate method, which are ideal for waters with little or no pollution. A distillation and titration approach is more suitable for wastewaters that contain high quantities of ammonia. The Kjeldahl method also includes the determination of total ammonia nitrogen (Kjeldahl 1883).

8.3.2   Nitrate and Nitrite

The most prevalent form of combined nitrogen that may be found in natural water-ways is the nitrate ion (NO^{3-}). By means of denitrification processes, mainly occurring in anaerobic environments, it may be biochemically reduced to nitrite (NO^{2-}). The nitrite ion is rapidly transformed into nitrate (Hakeem et al. 2011; Ozturk et al. 2013). Igneous rocks, land drainage, and plant and animal waste are some of the natural sources of nitrate in surface waterways. For aquatic plants, nitrate is an essential nutrient, and seasonal variations can be brought on by plant development and decay. Municipal and industrial wastewaters, particularly leachates from waste disposal sites and sanitary landfills, may increase naturally occurring concentrations of NO^{3-}-N, otherwise, whose natural concentrations rarely surpass 0.1 mg l^{-1}. The usage of inorganic nitrate fertilizers can be a substantial source in rural and suburban settings. Surface waters can have up to 5 mg of nitrate per liter when influenced by human activity, but they frequently have concentrations of less than 1 mg l^{-1}. Concentrations over 5 mg l^{-1} NO^{3-}-N are typically indicative of fertilizer runoff or human or animal waste pollution. Concentrations of NO3-N could exceed 200 mg l^{-1} in circumstances of severe pollution. The World Health Organization (WHO) recommends a 50 mg l^{-1} (or 11.3 mg l^{-1} as NO^{3-}-N) maximum limit for NO3- in drinking water, and waters with higher concentrations can pose a serious health concern (WHO 1985; Hakeem et al. 2011; Ozturk et al. 2013). Nitrate concentrations above 0.2 mgl^{-1} NO^{3-}-N in lakes tend to boost algae development and signal potential eutrophic conditions.

Nitrate naturally exists in groundwater due to soil leaching, but in locations where nitrogen fertilizer is applied in large quantities, it can accumulate to very high concentrations (500 mg l^{-1} NO^{3-}-N) (Ozturk et al. 2013). Increased fertilizer applications, notably in many traditional agricultural regions of Europe (Roberts and Marsh 1987) and India (Pali district of Rajasthan), and dramatic rises in nitrate concentrations in groundwater over the past six decades have been reported. However, nitrate leaching to groundwater is not exclusively caused by increased fertilizer use. Unfertilized grassland and natural plants rarely leak nitrate, but the soils in these locations have enough organic matter in them to represent a significant potential source of nitrate (due to the activity of nitrifying bacteria in the soil). The enhanced soil aeration that results from clearing and ploughing for cultivation promotes nitrifying bacterial activity and soil nitrate generation.

Nitrate and nitrite concentrations in surface waters provide a general indication of the extent of organic contamination and nutritional condition. As a result, these species are typically included in multipurpose or background monitoring programs, basic water quality assessments, and programs that particularly track the effects of relevant industrial or organic inputs (Hedayati and Sargolzaei 2013). Due to the potential health danger posed by excessive nitrate concentrations, it is also tested in sources of drinking water. When nitrate concentrations in the source water are high, the treated drinking water should also be analyzed because little nitrate is removed during the regular operations for treating drinking water. Nitrate and/or nitrite samples should be collected in glass or polyethylene vials, quickly filtered, and then analyzed. If this is not practicable, the sample can be delayed from decomposing by bacteria by adding 2–4 ml of chloroform per liter. The sample can be chilled and then kept at 3–4 °C for storage. Due to interferences from other compounds in the water that make nitrate identification difficult, the precise method selection may change depending on the anticipated nitrate concentration as N. Alternative methods include chemically analyzing one component of the sample for total inorganic nitrogen and the other for nitrite, then determining the nitrate concentration from the difference between the two results. Spectrophotometric techniques can be used to calculate nitrite concentrations. Colorimetric comparator methods, which are sold as kits, can be used to make some quick field determinations with a limited degree of precision.

8.3.3   Organic Nitrogen

Protein components, including as amino acids, nucleic acids, and urine, as well as the byproducts of their metabolic transformations make up the majority of organic nitrogen (e.g., humic acids and fulvic acids). Since organic nitrogen is primarily generated in water by bacteria and phytoplankton and then circulated through the food chain, it is naturally influenced by the seasonal changes in the biological community. Increased levels of organic nitrogen may be a sign of water body contamination. The Kjeldahl method, which yields total ammonia nitrogen + total organic nitrogen, is typically used to calculate organic nitrogen (Kjeldahl Nitrogen). The total organic nitrogen content is calculated as the difference between total nitrogen and its inorganic forms. Since organic nitrogen quickly turns into ammonia, samples must be unfiltered and analyzed within 24 h. If necessary, this procedure can be slowed down by adding 2–4 ml of chloroform or roughly 0.8 ml of concentrated H_2SO_4 per liter of sample. When this is necessary, the condition and length of preservation should be reported with the results. Storage should be done between 2 and 4 °C. The Kjeldahl method can alternatively be replaced by photochemical techniques. These procedures convert all organic nitrogen (including ammonia) to nitrates and nitrites; therefore measurements of these substances must have been made on the sample in advance. Total organic nitrogen is measured rather than total dissolved nitrogen if samples are screened.

8.4   Phosphorus Compounds

For living things to survive, phosphorus is a necessary nutrient. It can be found in water bodies as both dissolved and particulate species. It often controls the principal productivity of a water body since it is the nutrient that limits algae development. The main factor contributing to eutrophication is unnatural augmentation in concentration brought on by human activity.

Phosphorus is mostly found as dissolved orthophosphates, polyphosphates, and organically bound phosphates in natural waters and wastewaters. Continuous transitions between these forms take place as a result of the synthesis and disintegration of forms that are organically bound and oxidized inorganic forms. It is advised to represent phosphate concentrations as phosphorus, such as mgl^{-1} $PO^{4-}P$ (and not as $mg\ l^{-1}$ PO_4^{3-}).

The weathering of apatite (phosphorus-bearing rocks) and the breakdown of organic matter are the main natural sources of phosphorus. Elevated levels in surface waters are caused by domestic wastewaters (especially those containing detergents), industrial effluents, and fertilizer runoff. Bacteria can also mobilize phosphorus associated with organic and mineral components of sediments in water bodies and release it into the water column.

Due to the active uptake of phosphorus by plants, significant amounts of phosphorus are rarely encountered in freshwaters. As a result, seasonal variations in concentrations in surface waters may be significant. Phosphorus concentrations in natural surface waters typically vary from 0.005 to 0.020 mg l^{-1} $PO^{4-}P$. In certain pristine waterways, concentrations of $PO^{4-}P$ can be as low as 0.001 mg l^{-1} and as high as 200 mg l^{-1} in enclosed saline waters. The average concentration of $PO^{4-}P$ in groundwater is 0.02 mg l^{-1}. Phosphorus is frequently included in fundamental water quality surveys or background monitoring programs since it is a crucial part of the biological cycle in water bodies. Phosphates are mostly responsible for eutrophic conditions and can indicate the existence of pollution. Understanding the phosphate levels is necessary for lake or reservoir management, especially when it comes to the provision of drinking water, as it can be used to interpret algae growth rates.

Generally, total inorganic phosphate, orthophosphates, or total phosphorus is used to measure phosphorus concentrations (organically combined phosphorus and all phosphates). After passing the sample through a pre-washed 0.45-μm-pore-diameter membrane filter, the dissolved forms of phosphorus are quantified. The difference between total and dissolved concentrations can be used to calculate particle concentrations. Because sample containers often adsorb phosphorus, it is recommended that they be properly rinsed with the sample before use. Chloroform can be used to preserve samples for phosphate analysis, and they can be kept at 2–4 °C for up to 24 hours. If 1 ml of 30% sulfuric acid is supplied for every 100 ml of sample, samples for total phosphorus analyses can be stored in a glass flask with a firmly fitting glass stopper. It is crucial that samples are filtered as soon as feasible after collection for dissolved phosphorus. Phosphorus is determined by converting it to orthophosphate, which is then determined colorimetrically.

8.5   Bio-monitoring

The term "biological monitoring" or "bio-monitoring" refers to the use of biological responses to evaluate environmental changes, usually those brought on by anthropogenic activity. Qualitative, semi-quantitative, or quantitative bio-monitoring programs are all possible (Aksoy and Ozturk 1996, 1997; Aksoy et al. 2000). Programs for various kinds of water quality monitoring are increasingly using the useful assessment tool known as "bio-monitoring" (Saras 2021). Indicators, indicator species, or indicator communities are used in bio-monitoring. Typically, fish, algae, or benthic macroinvertebrates are employed. Additionally, some aquatic plants have been utilized as markers for contaminants like nitrogen enrichment (Alam and Hoque 2018). Environmental circumstances are reflected in the presence or absence of an indicator, an indicator species, or an indicator community. The lack of a species may not be as significant as it first appears because it may exist for causes other than pollution (e.g., predation, competition, or geographic barriers which prevented it from ever being at the site). More evidence of pollution than the loss of a single species can be found in the absence of several species of various orders with similar tolerance levels that were previously present at the same site. Which species should be present at the site or in the system must be known. Bacterial indicators, for example, *Salmonella typhimurium* and *Clostridium* sp. (Kalkan and Altuğ 2015), fecal coliforms (Saber et al. 2015), microbial indicators (Khatri and Tyagi 2015), planktonic indicators (Thakur et al. 2013), and fungal indicators (Hasselbach et al. 2005) are various bio-indicators for studying aquatic nutrient pollution. Algae have been extensively studied as potential bio-indicators of pollution (Hosmani 2013; Khalil et al. 2021). The amounts of nutrients and sunlight are both necessary for algal growth. Algae overgrowth is a sign of cultural eutrophication due to nutrient pollution (Zaghloul et al. 2020). Additionally, certain pollutants may not immediately harm other organisms clearly or may affect other populations at higher quantities because algae are sensitive to them (Hosmani 2013; Khalil et al. 2021).

References

Aksoy A, Ozturk M (1996) *Phoenix dactylifera* as a biomonitor of heavy metal pollution in Turkey. J Trace Microprobe Tech 14(3):605–614

Aksoy A, Ozturk M (1997) *Nerium oleander* as a biomonitor of lead and other heavy metal pollution in mediterranean environments. Sci Total Environ 205:145–150

Aksoy A, Celik A, Ozturk M, Tulu M (2000) Roadside plants as possible indicators of heavy metal pollution in Turkey. Chemia I Inzynieria Ekologiczna 7(11):1152–1162

Alam AR, Hoque S (2018) Phytoremediation of industrial wastewater by culturing aquatic macrophytes, *Trapa natans* L. and *Salvinia cucullata* Roxb. Jahangirnagar University. J Biological Sciences 6(2):19–27. https://doi.org/10.3329/jujbs.v6i2.36587

Hakeem KR, Ahmad A, Iqbal M, Gucel S, Ozturk M (2011) Nitrogen efficient rice genotype can reduce nitrate pollution. Environ Sci Pollut Res 18:1184–1193

Hedayati A, Sargolzaei J (2013) A review over diverse methods used in nitrogen removal from wastewater. Recent Patents on Chemical Engineering 6. https://doi.org/10.2174/2211334711306020007

Hosmani SP (2013) Fresh aquatic algae as indicators of aquatic quality. Univers J Environ Res Technol 3:473–482

Kalkan S, Altuğ G (2015) Bio-indicator bacteria & environmental variables of the coastal zones: the example of the Güllük Bay, Aegean Sea, Turkey. Mar Pollut Bull 15(95):380–384

Khalil S, Mahnashi MH, Hussain M, Zafar N, Waqar-Un-Nisa KFS, Afzal U, Shah GM, Niazi
    UM, Awais M, Irfan M (2021) Exploration and determination of algal role as Bioindicator to
    evaluate water quality – Probing fresh water algae. Saudi J Biol Sci 28(10):5728–5737. https://
    doi.org/10.1016/j.sjbs.2021.06.004
Khatri N, Tyagi S (2015) Influences of natural and anthropogenic factors on surface and ground
    aquatic quality in rural and urban areas. Front Life Sci 8:23–39
Kjeldahl J (1883) A new method for the determination of nitrogen in organic matter. Zeitschrift für
    Analytische Chemie, Scientific Research Publishing 22:366–382
Ozturk M, Altay V (2018) Impact of climate change on sea level, bioresources, biodiversity
    marine invasive species, ecology and food web: Past, present, and future. In: Understanding
    the Problems of Inland Waters: Case Study for the Caspian Basin (UPCB), pp. 30–34, 12–14
    May 2018, Bakü, Azerbaijan
Ozturk M, Secmen O, Leblebici E (1996) Plants and pollutants in the Eber Lake. Ekoloji 20:14–16
Ozturk M, Gucel S, Sakcali S, Baslar S (2013) Nitrate and edible plants in the Mediterranean
    region of Turkey: an overview. In: Umar S et al (eds) Nitrate in leafy vegetables-toxicity and
    safety measures. I.K. International Publishing House Pvt Ltd, New Delhi-Bangalore, pp 17–51
Poikane S, Várbíró G, Kelly MG, Birk S, Phillips G (2021) Estimating river nutrient concentrations
    consistent with good ecological condition: more stringent nutrient thresholds needed. Ecol
    Indicat 121:107017
Qin B, Yang L, Chen F, Zhu G, Zhang L, Chen Y (2006) Mechanism and control of lake eutrophi-
    cation. Chinese Sci Bull 51:2401–2412. https://doi.org/10.1007/s11434-006-2096-y
Roberts G, Marsh T (1987) The effects of agricultural practices on the nitrate concentrations in
    the surface water domestic supply sources of Western Europe. International Association of
    Hydrological Sciences Publication 164:365–380
Saber M, Abouziena HF, Hoballah E, El-Ashry S, Zaghloul AM (2015) Phytoremediation of
    potential toxic elements in contaminated sewaged soils by Sunflower (Helianthus annuus) and
    Corn (Zea mays L.) plants. In: 12th International Phytotechnology Conference, Manhattan
Saras DR (2021) Bio-monitoring is the most important tool for the assessment of the pollution in
    the lentic water systems of Kanpur (Dehat). http://www.jetir.org/, Available at SSRN: https://
    ssrn.com/abstract=3918501 or https://doi.org/10.2139/ssrn.3918501
Thakur RK, Jindal R, Singh UB, Ahluwalia AS (2013) Plankton diversity and aquatic quality
    assessment of three fresh aquatic lakes of Mandi (Himachal Pradesh, India) with special refer-
    ence to planktonic indicators. Environ Monit Assess 185:8355–8373
WHO (1985) Health hazards from nitrate in drinking-water. Report on a WHO meeting,
    Copenhagen, 5–9 March 1984. Copenhagen, WHO Regional Office for Europe (Environmental
    Health Series No. 1)
Zaghloul A, Saber M, Gadow S, Awad F (2020) Biological indicators for pollution detection
    in terrestrial and aquatic ecosystems. Bull Natl Res Cent 44:127. https://doi.org/10.1186/
    s42269-020-00385-x

Chapter 9
Impacts of Nutrient Pollution

9.1 Introduction

Nutrient pollution refers to overloading of aquatic environs by excessive influx of nutrients. Every creature has to taste death. The aging of surface waters is an inevitable fact. The life cycle of aquatic environs surpasses oligotrophic, mesotrophic, eutrophic, and dystrophic stage; each stage proceeds at a very slow rate (Ozturk et al. 2005). But due to anthropogenic intervention, the rate of aging of water body is accelerated by agricultural runoff, water-induced soil erosion, sewage dumping, excessive use of detergents and leakages from septic tanks, latrine, and kitchen wastes (Yucel et al. 1995; Cetin et al. 2000; Soker et al. 2006; Sakcali et al. 2009; Hakeem et al. 2011; Ozturk et al. 2011). Being highly enriched with nutrients (nitrogen and phosphorous), sewage dumping is the noxious practice in various parts of the world. Influx of nutrient-loaded sewage or runoff contributes to nutrient enrichment of water bodies known as accelerated eutrophication. Accelerated eutrophication is also known as cultural eutrophication or anthropogenic eutrophication which results in proliferation of weeds and algal blooms (Fig. 9.1). The nutrient enrichment of water body causes deteriorated water quality due to synthetic as well as organic fertilizers, anoxic conditions, aluminum toxicity, and fecal matter-related diseases which negatively impact fish and other sensitive organisms including human beings. Each of the impacts of nutrient pollution is briefly discussed in this chapter.

© The Author(s), under exclusive license to Springer Nature Switzerland AG 2023
M. A. Dervash et al., *Phytosequestration*, SpringerBriefs in Environmental
Science, https://doi.org/10.1007/978-3-031-26921-9_9

Fig. 9.1 Proliferation of aquatic weeds following excessive nutrient loads (eutrophication)

9.2   Sources

The sources of nutrient pollution may emerge as a point source and nonpoint or diffused sources. Nonpoint sources are more difficult to identify. The main sources include fertilizer-laden agricultural runoff, wastes from farmyard, stormwater runoff from urban as suburban areas, excessive fertilizer use in lawns, golf courses and public parks, municipal solid waste, industrial effluents, and raw sewage dumping (Tao et al. 2021; Berger et al. 2022). Air pollution can also owe independently to nutrient pollution after long-range transportation and settling of air pollutants especially nitrogen from remote sources.

9.3   Oxygen Depletion

When organic and synthetic fertilizers find their way through agricultural runoff into surface water, the growth of microorganisms is triggered. Microorganisms require dissolved oxygen of an aquatic setup to breakdown the nutrients which in turn support their growth and survival. In due practice, dissolved oxygen in an aquatic setup is depreciated which negatively impacts survival of fish and other aquatic biotics by pushing them to suffocation. Later the hypoxic or anoxic conditions result in the death of aquatic biota by formation of dead zones (Knockaert 2022). There is change in physicochemical characteristics of water which further cause acidification of water body. As biological oxygen demand escalates, dissolved oxygen decelerates which results in generation of unpleasant odors.

9.4 Weed Growth and Algal Blooms

As the nutrient supply is increased into an aquatic body, there is proliferation of aquatic weeds and algal blooms (Turkan et al. 1989). As nutrient availability is the principal factor for growth and development of aquatic plants, its shortage may limit the primary productivity in an aquatic system. When nutrients are present in abundance, aquatic plants flourish which is not an indication of a healthy water body. Microorganisms depend upon a dead organic matter to continue their life cycle which ultimately causes dissolved oxygen deficit water column and takes toll of aquatic biota. Additionally, toxicity hazard of algae poses threat to human health (Sha et al. 2021). Occurrence of the neurotoxin BMAA (beta-N-methylamino-L-alanine) in cyanobacterial blooms has been implicated to cause myriad health implications (Violi et al. 2019) predominating to nausea, respiratory ailments, Alzheimer's disease, Parkinson's disease, and amyotrophic lateral sclerosis in humans. It has also been documented to be detrimental to lifestock.

9.5 Ammonia Toxicity

Ammonia-laden agricultural runoff from farmyard manure croplands is lethal to aquatic life. At elevated doses in surface water, ammonia is notoriously known to kill fish (by damaging their gills) and may negatively affect species diversity, thus detrimental to aquatic ecosystems (Bhakta 2006). As fishes are very sensitive to ammonia concentrations, levels of even 0.02 parts per million may be toxic at pH 5. Increased aluminum consumption with drinking water can cause osteomalacia in humans (which is marked by calcium and vitamin D deficiency).

9.6 Fecal Matter Contamination

The fresh farmyard manure possesses numerous microorganisms. Some of which are pathogenic in nature. The agricultural runoff, direct dumping of raw sewage, and blackwater (wastewater from bathrooms and toilets containing fecal matter and urine) are the main sources of fecal matter contamination in surface water resources (Reynolds et al. 2021). Thus, coliform bacteria count acts as an indicator of water pollution. *Escherichia coli* is a strong indicator of sewage contamination. The presence of *E. coli* in drinking water may cause many health issues in children and elderly people by producing a toxin known as "Shiga" which destroys the intestinal lining in humans (Reynolds et al. 2021).

9.7 Nitrate Toxicity

Excessive nitrate in agricultural soils is removed through leaching and runoff (Hakeem et al. 2011; Ozturk et al. 2005, 2011, 2013). Nitrate pollution is a notorious type of water pollution which is the culprit behind nitrate poisoning in humans. Normally, oxygen binds with hemoglobin and makes oxyhemoglobin which is an essential phenomenon for liberation of energy at the cellular level. Excessive levels of nitrate in drinking water interfere with oxygen uptake in our blood and lead to the formation of methemoglobin instead of oxyhemoglobin. The formation of methemoglobin is known to cause a "blue baby syndrome" (methemoglobinemia) in human infants. Above permissible limits, nitrates and nitrites in drinking water can increase risk of gut cancers (Picetti et al. 2022). The effects of nitrate pollution on two freshwater mussels (*Lampsilis siliquoidea* and *L. fasciola*) show decreased production in a number of mussels, decelerated glochidia attachment, and metamorphosis activities (Moore and Bringolf 2018).

References

Berger M, Canty SWJ, Tuholske C, Halpern BS (2022) Sources and discharge of nitrogen pollution from agriculture and wastewater in the Mesoamerican Reef region. Ocean Coast Manage 227:106269. https://doi.org/10.1016/j.ocecoaman.2022.106269

Bhakta J (2006) Ammonia toxicity to four freshwater fish species: Catla catla, Labeo bata, Cyprinus carpio and Oreochromis mossambica. Electron J Biol 2:39–41

Cetin E, Unlu Y, Ozturk M, Tulu M (2000) Trace element studies on Halilurrahman Lake Sanlıurfa-Turkey. Chemia I Inzynieria Ekologiczna 7(11):1143–1151

Hakeem KR, Ahmad A, Iqbal M, Gucel S, Ozturk M (2011) Nitrogen efficient rice genotype can reduce nitrate pollution. Environ Sci Pollut Res 18:1184–1193

Knockaert C (2022) Possible consequences of eutrophication. Available from http://www.coastal-wiki.org/wiki/Possible_consequences_of_eutrophication. Accessed 17 Nov 2022

Moore AP, Bringolf RB (2018) Effects of nitrate on freshwater mussel glochidia attachment and metamorphosis success to the juvenile stage. Environ Pollut 242(Pt A):807–813. https://doi.org/10.1016/j.envpol.2018.07.047. PMID: 30032077

Ozturk M, Alyanak I, Sakcali S, Guvensen A (2005) Multipurpose plant systems for renovation of waste waters. Arab J Sci Eng 30(2C):17–28

Ozturk M, Gucel S, Sakcali S, Guvensen A (2011) An overview of the possiblities for wastewater utilization for agriculture in Turkey. Israel J Plant Sci 59:223–234

Ozturk M, Gucel S, Sakcali S, Baslar S (2013) Nitrate and edible plants in the Mediterranean Region of Turkey: an overview. In: Umar S et al (eds) Nitrate in leafy vegetables-toxicity and safety measures. I.K. International Publishing House Pvt. Ltd., New Delhi-Bangalore, pp 17–51

Picetti R, Deeney M, Pastorino S, Miller MR, Shah A, Leon DA, Dangour AD, Green R (2022) Nitrate and nitrite contamination in drinking water and cancer risk: a systematic review with meta-analysis. Environ Res 210:112988. https://doi.org/10.1016/j.envres.2022.112988. PMID: 35217009

Reynolds LJ, Martin NA, Sala-Comorera L, Callanan K, Doyle P, O'Leary C, Buggy P, Nolan TM, O'Hare GMP, O'Sullivan JJ, Meijer WG (2021) Identifying Sources of Faecal Contamination in a Small Urban Stream Catchment: A Multiparametric Approach. Front Microbiol 12:661954. https://doi.org/10.3389/fmicb.2021.661954

Sakcali S, Yilmaz R, Gucel S, Yarci C, Ozturk M (2009) Water pollution studies in the Rivers of Edirne State–Turkey. Aquat Ecosyst Health Manage 12(3):313–319

Seker S, Ileri R, Ozturk M (2006) Evaluation of activated sludge by white rot fungi for decolorization of textile waste waters. J World Assoc Soil Water Conserv 1(7):81–87

Sha J, Xiong H, Li C, Lu Z, Zhang J, Zhong H, Zhang W, Yan B (2021) Harmful algal blooms and their eco-environmental indication. Chemosphere 274:129912, ISSN 0045-6535. https://doi.org/10.1016/j.chemosphere.2021.129912

Tao W, Niu L, Dong Y, Fu T, Lou Q (2021) Nutrient pollution and its dynamic source-sink pattern in the Pearl River Estuary (South China). Front Mar Sci 8:713907. https://doi.org/10.3389/fmars.2021.713907

Turkan I, Sukatar A, Ozturk M (1989) Heavy metal accumulation by the algae in the bay of Izmir. Rev Int Oceanogr Med 93:71–76

Violi JP, Facey JA, Mitrovic SM, Colville A, Rodgers KJ (2019) Production of β-methylamino-L-alanine (BMAA) and its isomers by freshwater diatoms. Toxins 11(9):512. https://doi.org/10.3390/toxins11090512

Yucel E, Dogan F, Ozturk M (1995) Heavy metal status of Porsuk stream in relation to public health. Ekoloji 17:29–32

Chapter 10
Phytoremediation of Nuisance Pollution

10.1 Introduction

Aquatic contamination due to nutrient enrichment is a worldwide concern which is an aftermath of domestic and agricultural runoff. Nowadays, in the present era of industrialization and urbanization, aquatic environs are heavily loaded with contaminants like nitrogen (N) and phosphorous (P) which lead to proliferation of macrophytes (Fig. 10.1).

Acidic precipitation, wastewaters from industrial (loaded with heavy metals) and human settlements (loaded with organic matter and detergents) (Cetin et al. 2000; Verla et al. 2018), and numerous other inorganic and organic agricultural compounds are the principal causes of water pollution. Also, different ionic forms of nitrogen and phosphorus cause the accelerated nutrient enrichment of inland waters (Cetin et al. 2000; Khan and Ansari 2005). Nutrient remediation, nutrient trade, and nutrient source allocation are some of the ways for reducing and managing nutrient contamination. There are chemical, physical, and biological techniques available for remediation of nutrients, but nutrient sequestration via plants is an ecofriendly approach which can be applied to get rid of nuisance contaminant. Therefore, in order to excess nutrients from the contaminated media, nutrient sequestration can be exploited.

10.2 Nutrient Trading

A market-based policy contrivance exploited to enhance or sustain water quality and nutrient trading is a subset of "water quality trading" (George et al. 2012). The idea of water quality trading is predicated on the assumption that controlling the same pollutant might have substantially varied costs depending on the pollution source within a watershed. The same concepts apply to trade in nutrient water

Fig. 10.1 Phytoremediation of nutrient pollution using *Trapa natans*

quality, which entails the voluntary transfer of pollution reduction credits from sources with low pollution control costs to those with high pollution control costs (George et al. 2012). The core idea is "polluter pays principle," which is typically connected to a regulatory prerequisite for taking part in the trading program.

10.3   Nutrient Source Appointment

After attenuation or treatment, the nutrient load from different inflowing sources into the aquatic environs is estimated using nutrient source apportionment. Nitrogen in water bodies is mainly contributed by agricultural runoff (Picetti et al. 2022), whereas phosphorous is contributed by excessive use of detergents in households and industries. There are two main strategies to load apportionment modeling: (a) source-oriented approaches, which calculate diffuse pollution emissions using models typically based on export coefficients from catchments with comparable characteristics, and (b) load-oriented approaches, which assign origin based on in-stream monitoring data (Greene et al. 2011).

10.4   Phytoremediation: A Plant-Based
Ecofriendly Technology

The phytoremediation refers to "the efficient use of plants to remove, detoxify or immobilize environmental contaminants" (Hakeem et al. 2011; Ozturk et al. 2005, 2011a, b; Seker et al. 2006; Ansari et al. 2020), which is a quite sustainable approach for cleaning up the polluted environs. The process is initiated in the rhizosphere of plants where from pollutants are absorbed, accumulated in plant tissues, and degraded or comparably transformed into environmentally less detrimental states. The technique of phytoremediation has been studied exhaustively during the last several decades (Favas et al. 2018), but actual evaluation of aquatic plants in wastewater treatment is more recent (Ozturk et al. 2004a, b, 2005, 2011a, b; 2015a, b; Carolin et al. 2017). It has yielded fruitful outcomes in present era also (Prasad 2006) by employing this ecofriendly strategy practically in the field across the world (Vidal et al. 2019; Yadav et al. 2018). The phytoremediation potential of plant species depends upon photosynthetic activity, growth rate, span of life cycle, and adaptability of a plant to withstand pollution load (Jamuna and Noorjahan 2009). Phytoremediation encompasses many plant-based technologies which includes filtration of contaminated water in the rhizosphere of plants which is similar to sieving technique (rhizofiltration), absorption, and degradation of contaminants within root zone (rhizodegradation) with the aid of microbial intervention, degradation of harmful pollutants to less detrimental states within plant tissues (phytodegradation), transformation of harmful instable forms to highly stable forms (phytostabilization), and volatilization of certain transformed volatile compounds sequestered from the contaminant load (phytovolatilization) (Girdhar et al. 2014) driven by the evapotranspiration of plants. In terrestrial plants, the rhizospheric zone possesses the ability to adsorb and demobilize the contaminants (phytostabilization) and even the plant has the potential to invest its extraction prospective through its harvestable biomass for various detrimental compounds like heavy metals and recalcitrant organic compounds (phytoextraction) (Ali et al. 2020), macrophytes possess specialized potential and structure that aid in nutrient absorption from a polluted aquatic environment. In comparison to other macrophytes, *Lemna* (duckweed) can withstand harsh environmental conditions with wide pH range tolerance and advanced biomass production rate (Ozturk et al. 1994). Phytoremediation potential of macrophytes has been extensively used in ecotoxicological studies (Chaudhary and Sharma 2014).

Eichhornia crassipes has been extensively adopted for nutrient sequestration of highly polluted water bodies with high ammonium nitrate and phosphate, ammonia, total nitrogen, increased biological oxygen demand, and total suspended solids due its high photosynthetic activity and more turnover rate (Ozturk et al. 2004a, b, 2005, 2011a, b, 2015a, b; Ekperusi et al. 2019). Similarly, *Pistia stratiotes, Lemna* sp., *Spirodela, Wolffia, Typha* sp., *Phragmites australis, Chrysopogon zizanioides, Salvinia natans,* and *Azolla* possess great phytosequestration potential under eutrophic conditions (Ansari et al. 2020). Aquatic plants exhibit the characteristic of altering the physicochemical parameters of water by photosynthetic activity which utilizes diffused sunlight, nutrients, and the dissolved carbon dioxide from water. And after the photolysis of water, oxygen is liberated by the plant which in turn increases dissolved oxygen content pushing up the pH of water (Obinna and Ebere 2019). The altered environment by macrophytic activity further favors capture and storage of nutrients in the form of biomass (nutrient sequestration/phytosequestration through phytoremediation).

References

Ali S, Abbas Z, Rizwan M, Zaheer IE, Yavaş İ, Ünay A, Abdel-Daim MM, Bin-Jumah M, Hasanuzzaman M, Kalderis D (2020) Application of floating aquatic plants in phytoremediation of heavy metals polluted water: a review. Sustainability 12(5):1927. https://doi.org/10.3390/su12051927

Ansari AA, Naeem M, Gill SS, AlZuaibr FM (2020) Phytoremediation of contaminated waters: an eco-friendly technology based on aquatic macrophytes application. Egypt J Aquat Res 46(4):371–376

Carolin CF, Kumar PS, Saravanan A, Joshiba GJ, Naushad M (2017) Efficient techniques for the removal of toxic heavy metals from aquatic environment: a review. J Environ Chem Eng, vol 5, pp 2782–2799

Cetin E, Unlu Y, Ozturk M, Tulu M (2000) Trace element studies on Halilurrahman Lake Sanlıurfa-Turkey. Chemia I Inzynieria Ekologiczna 7(11):1143–1151

Chaudhary E, Sharma P (2014) Duckweed plant: a better future option for phytoremediation. Int J Emerg Sci Eng 2:39–41

Ekperusi AO, Sikoki FD, Nwachukwu EO (2019) Application of common duckweed (*Lemna minor*) in phytoremediation of chemicals in the environment: state and future perspective. Chemosphere 223:285. https://doi.org/10.1016/j.chemosphere.02.025

Favas PJC, Pratas J, Rodrigues N, D'Souza R, Varun M, Paul MS (2018) Metal (loid) accumulation in aquatic plants of a mining area: potential for water quality biomonitoring and biogeochemical prospecting. Chemosphere 194:158–170

George VH, Ross L, Sarah C, Robert B, Justin B (2012) Nutrient credit trading for the Chesapeake Bay: an economic study

Girdhar M, Sharma NR, Rehman H, Kumar A, Mohan A (2014) Comparative assessment for hyperaccumulatory and phytoremediation capability of three wild weeds. Biotech 4:579–589

Greene S, Taylor D, McElarney YR, Foy RH, Jordan P (2011) An evaluation of catchment-scale phosphorus mitigation using load apportionment modelling. Science of The Total Environment 409(11):2211–2221. https://doi.org/10.1016/j.scitotenv.2011.02.016

Hakeem KR, Ahmad A, Iqbal M, Gucel S, Ozturk M (2011) Nitrogen efficient rice genotype can reduce nitrate pollution. Environ Sci Pollut Res 18:1184–1193

Jamuna S, Noorjahan CM (2009) Treatment of sewage waste water using water hyacinth – Eichhornia sp. and its reuse for fish culture. Toxico Int 16(2):103–106

Khan FA, Ansari AA (2005) Eutrophication: An Ecological Vision. The Botanical Review 71:449–482. https://doi.org/10.1663/0006-8101(2005)071

Obinna IB, Ebere EC (2019) Phytoremediation of polluted water bodies with aquatic plants: Recent Progress on heavy metal and organic pollutants. Anal Methods Environ Chem J 10.20944/preprints201909.0020.v1

Ozturk M, Uysal T, Guvensen A (1994) *Lemna minor* L. as a Water Cleaner. XII. National Biology Congress, Edirne, pp 68–70

Ozturk M, Alyanak I, Uysal I, Guvensen A, Sakcali S (2004a) Multipurpose Plant Systems for renovation of wastewaters. International Seminar on Salinity Mitigation for Efficient Water Resources Management Proceedings No.11, CEWRE Publ. 194, Lahore-Pakistan, 160-170.39

Ozturk M, Alyanak I, Guvensen A (2004b) Renovation of wastewaters using flowering plants. In: First International Symposium on Environment Protection for Sustainable Development & Biotechnological Aspects, 15–17 December, Fez- Morocco

Ozturk M, Alyanak I, Sakcali S, Guvensen A (2005) Multipurpose plant systems for renovation of waste waters. Arab J Sci Eng 30(2C):17–28

Ozturk M, Gucel S, Sakcali S, Guvensen A (2011a) An overview of the possibilities for wastewater utilization for agriculture in Turkey. Israel J Plant Sci 59:223–234

Ozturk M, Mermut A, Celik A (2011b) Land degradation, urbanisation, land use & environment, NAM S. & T. (Delhi-India), 445 pp.

Ozturk M, Ashraf M, Aksoy A, Ahmad MSA (2015a) Phytoremediation for green energy. Springer Science+Business Media, New York, 191 pp

Ozturk M, Ashraf M, Aksoy A, Ahmad MSA (2015b) Plants, Pollutants & Remediation. Springer Science+Business Media, New York, 407 pp

Picetti R, Deeney M, Pastorino S, Miller MR, Shah A, Leon DA, Dangour AD, Green R (2022) Nitrate and nitrite contamination in drinking water and cancer risk: A systematic review with meta-analysis. Environ Res 210:112988. https://doi.org/10.1016/j.envres.2022.112988. PMID: 35217009

Prasad MNV (2006) Aquatic plants for phytotechnology. In: Singh SN, Tripathi RD (eds) Environmental bioremediation technologies. Springer, Berlin, Heidelberg, pp 259–274

Seker S, Ileri R, Ozturk M (2006) Evaluation of activated sludge by white rot fungi for decolorization of textile waste waters. J World Assoc Soil Water Conserv 1(7):81–87

Verla AW, Verla EN, Amaobi CE, Enyoh CE (2018) Water pollution scenario at River Uramurukwa flowing through Owerri Metropolis, Imo State, Nigeria. Int J Sci Res 3:40–46

Vidal CF, Oliveira JA, da Silva AA, Ribeiro C, Farnese FDS (2019) Phytoremediation of arsenite-contaminated environments: Is Pistia stratiotes L. a useful tool? Ecol Indicat 104:794–801

Yadav KK, Gupta N, Kumar A, Reecec LM, Singh N, Rezania S, Khan SA (2018) Mechanistic understanding and holistic approach of phytoremediation: a review on application and future prospects. Ecol Eng 120:274–298

Index

© The Author(s), under exclusive license to Springer Nature Switzerland AG 2023
M. A. Dervash et al., *Phytosequestration*, SpringerBriefs in Environmental
Science, https://doi.org/10.1007/978-3-031-26921-9

Printed in the United States
by Baker & Taylor Publisher Services